Introduction to Control Theory

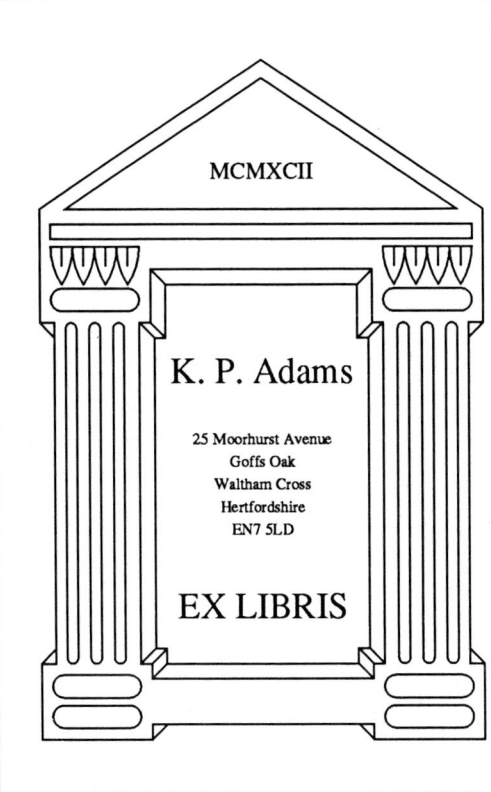

MCMXCII

K. P. Adams

25 Moorhurst Avenue
Goffs Oak
Waltham Cross
Hertfordshire
EN7 5LD

EX LIBRIS

Also from The Macmillan Press

Introduction to State-variable Analysis
  P. F. Blackman

Control Engineering, Theory, worked examples and problems
  R. V. Buckley

Fluid Power Systems, Theory, worked examples and problems
  A. B. Goodwin

Hydraulic Systems Analysis
  J. D. Stringer

Introduction to Control Theory

S. A. MARSHALL

*Chairman and Professor of Mechanical Engineering,
University of Wollongong,
New South Wales*

© S. A. Marshall 1978

All rights reserved. No part of this publication may be reproduced or transmitted, in any form or by any means, without permission.

First edition 1978
Reprinted 1981, 1983

Published by
THE MACMILLAN PRESS LTD
London and Basingstoke
Companies and representatives throughout
the world

Typeset in 10/11 IBM Press New Roman by
Reproduction Drawings Ltd, Sutton, Surrey

Printed in Hong Kong

British Library Cataloguing in Publication Data

Marshall, S. A.
Introduction to control theory
1. Control theory
I. Title
629.8'312   QA402.3

ISBN 0-333-18311-8
ISBN 0-333-18312-6 Pbk

The paperback edition of this book is sold subject to the condition that it shall not, by way of trade or otherwise, be lent, resold, hired out, or otherwise circulated without the publisher's prior consent in any form of binding or cover other than that in which it is published and without a similar condition including this condition being imposed on the subsequent purchaser.

To my wife, Margaret Rose

Contents

Preface   xi

1 Introduction   1

    1.1 Open- and Closed-loop Concepts   1
    1.2 Historical Development   2
    1.3 Outline of Text   2

2 Mathematical Modelling   4

    2.1 The Need for Mathematical Modelling   4
    2.2 Description of Simple Mechanical Systems   5
    2.3 More about Mathematical Modelling   9
    2.4 Model of a Trailer Suspension System   9
    2.5 Modelling of Simple Electrical Systems   12
    2.6 Modelling of Chemical Systems   16
    2.7 Linearisation of Non-linear Equations   18
    2.8 Concluding Remarks   20

3 The Need for Some Mathematics   23

    3.1 Classical Method of Solution of Linear Differential Equations   24
    3.2 Introduction to the Laplace Transformation   27
    3.3 Some Properties of the Laplace Transformation   28
    3.4 Laplace Transformations of Some Time Functions   30
    3.5 Roots of Polynomials   32
    3.6 Partial-fraction Expansions   35
    3.7 Solution of Linear Differential Equations Using the Laplace Transformation   39
    3.8 Second-order Differential Equations   41

Contents

4 **Control-system Representation** — 49

    4.1 Open-loop and Closed-loop Control — 49
    4.2 Transfer Functions — 53
    4.3 Block-diagram Representation — 57
    4.4 Signal-flow Graphs — 64
    4.5 Concluding Remarks — 69

5 **Steady-state and Transient Behaviour of Control Systems** — 73

    5.1 Performance Criteria — 73
    5.2 Absolute Stability: the Method of Routh — 75
    5.3 Steady-state Analysis and Classification of Systems — 79
    5.4 Transient Behaviour of Control Systems — 87
    5.5 Sensitivity of Control Systems — 96
    5.6 Behaviour of a Governor System — 98
    5.7 Use of Computer Graphics — 101

6 **The Root-locus Method** — 105

    6.1 Pole–Zero Configuration — 106
    6.2 Root-locus Method — 109
    6.3 Rules for Drawing the Root-locus Diagram — 111
    6.4 Determining the Transient Response — 120
    6.5 Use of Digital Computer and Graphics Terminal — 124

7 **Frequency-response Methods** — 129

    7.1 Derivation of the Frequency Response of a System — 130
    7.2 Graphical Representation of the Frequency-response Function — 136
    7.3 Stability from the Nyquist Diagram — 154
    7.4 Relative Stability: Gain and Phase Margins — 161
    7.5 Closed-loop Performance from Open-loop Frequency Characteristics — 16:
    7.6 Performance Requirements — 171
    7.7 System Sensitivity — 172
    7.8 Inverse Nyquist Diagram — 172

8 **Control-system Synthesis** — 175

    8.1 Series Compensation — 177
    8.2 Parallel Compensation — 184
    8.3 System Synthesis in the s-plane Using the Root-locus Method — 186
    8.4 System Synthesis in the $G(j\omega)H(j\omega)$ Plane Using the Frequency-response Method — 192
    8.5 System Synthesis in the $G(j\omega)H(j\omega)$ Plane Using the Nyquist Diagram — 198
    8.6 System Synthesis in the $G(j\omega)H(j\omega)$ Plane Using the Inverse Nyquist Diagram — 203
    8.7 Feedforward Control — 205
    8.8 Concluding Remarks — 207

9 Bridging the Gap — 210

    9.1 Representation of Multivariable Systems — 212
    9.2 The Inverse-Nyquist-array Technique — 216
    9.3 Conclusion — 224

Appendix — 226

Index — 229

Preface

This book has arisen, after many years' association with automatic control both in the university and in industry, from the growing conviction that there is a need for a new and modern text dealing with classical linear control theory, especially since many universities and polytechnics now have access to digital computers that have graphics terminals for interactive design work. Most of the classical control work is concerned with the analysis of the behaviour of closed-loop systems using knowledge of the open-loop system together with well-tried graphical techniques. It is this dependence that creates new possibilities for the use of control theory—there is now no need for laborious hand calculation and graph plotting since it is possible to automate all the techniques and carry out the analysis and synthesis of control systems via a graphics terminal with accuracy and speed. Further, it is possible to interact with the computing system and hence to build up progressive design techniques.

It has been recognised for some years that classical control theory has not been sufficiently developed to include techniques for designing industrial control systems in which there are many inputs and outputs and interaction between the control loops; hence alternative methods have been developed. Recent developments (at U.M.I.S.T.) have shown that the design of such control systems may be tackled in two parts: firstly, the interaction is systematically reduced until it can be safely ignored; and secondly, the many control loops may subsequently be designed using classical control theory. These developments in control theory stress the need for a thorough understanding of the classical approach. Indeed, in his book *Computer-Aided Control System Design*, Professor H. H. Rosenbrock states that 'The second development which will have a profound and increasing influence on all engineering design, is the availability of computers with graphical display. This allows a complete re-evaluation of the existing frequency-response methods, so that the graphical features are used only for communication with the designer, and not to replace standard numerical procedures for root solving etc. . . .'

This book was written mainly during my leisure hours and consequently was made possible only by the continued encouragement and understanding of my wife.

Wollongong, N.S.W.   S. A. MARSHALL

1 Introduction

1.1 OPEN- AND CLOSED-LOOP CONCEPTS

A central-heating system or air-conditioning system if operated in conjunction with a room thermostat is said to be automatically controlled, whilst if it is operating without the thermostat it is said to be manually controlled. These systems are shown diagrammatically in figures 1.1a and b, respectively. In the case of figure 1.1b, if a change should occur in the outside air temperature, a change in room temperature would result and manual intervention would be needed to correct the room temperature. For the automatically controlled system of figure 1.1a the effect of a change in outside air temperature on the room temperature would be corrected without manual intervention.

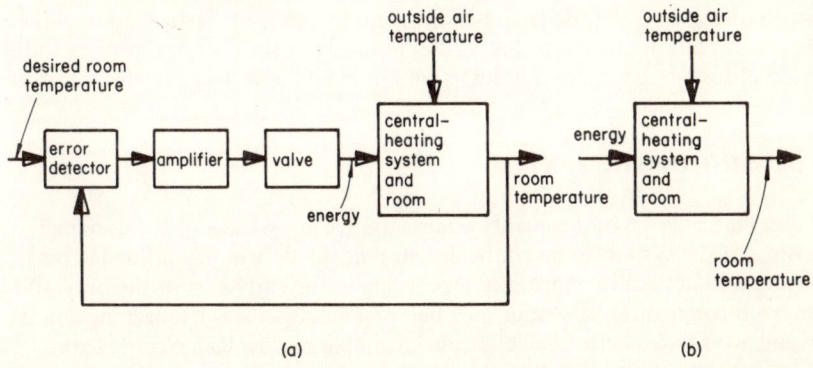

Figure 1.1

The manual system is said to operate in open loop and is known as an *open-loop system* whilst the automatic system is known as a *closed-loop system*. In a closed-loop system the actual output (room temperature) is measured and compared with the desired output (desired room temperature) and the difference, after amplification, is used to operate a valve or similar device (to increase or decrease the energy input) such that the difference is made as small as possible.

This type of closed-loop system or feedback system, which has to maintain an output equal to a desired value despite outside fluctuations or disturbances, is known as a *regulator*. On the other hand, a closed-loop system that has to produce an output position equal to some reference input position—as in the case of a gun-positioning system, which is not influenced by outside disturbances—is known as a *servomechanism*.

1.2 HISTORICAL DEVELOPMENT

There are many instances in early history of closed-loop systems, for example, the South-pointing chariot used by the early Chinese and the fantail mechanism of windmills, which ensures that the windmill is facing into the wind at all times. But the best-known application of a control system is that given by the flyball governor system, which is used to control the speed of a steam engine by opening and closing the governor valve. This device was first used before the end of the eighteenth century and is still used extensively today to control turbine speed in modern boiler-turbine units.

The early development of servomechanisms was by Minorsky (1922), who was concerned with the automatic steering of ships. This principle was also later applied to the positioning of shipboard guns, although the term servomechanism was first used by Hazen (1934) in his work dealing with position-control systems.

The development of the mathematical theory of feedback systems was first initiated by Nyquist (1932), who considered the response or behaviour of open-loop systems to sinusoidal variations and from it deduced the stability of the associated closed-loop system. This work was further developed by Bode (1938), who considered the frequency–phase characteristics of open-loop systems, and later by Evans (1948) with the introduction of his so-called root-locus method of design. The methods of Nyquist, Bode and Evans form the basis of classical control theory.

At present automatic control is used extensively in the chemical-process industry, the steel industry, the aviation industry and in manufacturing generally.

1.3 OUTLINE OF THE TEXT

The systematic design of a control system requires knowledge of the dynamic behaviour of the system to be controlled so that the design may influence that behaviour and hopefully improve it. Accordingly, the early part of the book outlines how this dynamic behaviour may be obtained, discusses its meaning and its form and gives insight into the behaviour of simple mechanical, electrical and chemical systems. Following this, it is necessary to present some fairly elementary mathematics—including the solution of linear differential equations, the Laplace transformation, roots of polynomials and the theory of algebraic equations—as a supplement to the mathematics usually contained in the first year of engineering degree and HND courses. Finally this section of the book gives an account of control-system representation and discusses in detail the behaviour of an industrial process, explaining many control terms with reference to this process. The derivation of transfer-function representation is amply illustrated.

Introduction

The main body of the book, which is concerned with the analysis of closed-loop control systems, is divided into three parts. The first part is concerned with the time behaviour of control systems and the effect of feedback and feedforward control and stability; the behaviour of a governor system is examined. The second part describes system pole–zero patterns and the use of the root-locus technique of Evans; this necessitates a discussion of a further branch of mathematics concerned with complex-variable theory. The determination of the transient response of the system is also considered, with an examination of how each closed-loop pole contributes to that response. The third part introduces frequency-response techniques and includes Bode, Nyquist and inverse Nyquist methods. The use of interactive graphics as an aid is stressed throughout.

Whereas the previous chapters have considered analysis, the penultimate chapter discusses the important aspects of closed-loop control-system synthesis, and includes the various compensation techniques available to the designer of control systems; full use is made of the root-locus and frequency-response methods and many illustrative examples are included. The final chapter discusses a method that may be said to bridge the gap between classical control theory and the modern theory mentioned earlier—the inverse Nyquist array technique—and should whet the student's appetite for further study.

The book will be suitable as an introduction to linear control theory for those in the second or final year of a degree course in electrical, mechanical or chemical engineering, for second-year control-engineering students and for some final-year HND students. It is amply illustrated and includes many worked examples and problems so that students can assess their progress as they work through the chapters; the problems are typical of those being set in papers for honours engineering students. The references have been carefully selected to amplify and extend particular topics.

REFERENCES

Bennett, S., 'The Search for "Uniform and Equable Motion". A Study of the Early Methods of Control of the Steam Engine', *Int. J. Control*, 21 (1975) 113.–'The Emergence of a Discipline: Automatic Control 1940–1960', *Automatica*, 12 (1976) 113.
Bode, H. W., *Amplifiers*, Patent 2 (1938) 123, 178.
Evans, W. R., 'Graphical Analysis of Control Systems', *Trans. A.I.E.E.*, 67 (1948) 547.
Hazen, H. L., 'Theory of Servomechanisms', *J. Franklin Inst.*, 218 (1934) 279.
Minorsky, N., 'Directional Stability of Automatically Steered Bodies', *J. Am. Soc. Nav. Engrs*, 34 (1922) 280.
Nyquist, H., 'Regeneration Theory', *Bell Syst. Tech. J.*, 13 (1932) 1.

2 Mathematical Modelling

2.1 THE NEED FOR MATHEMATICAL MODELLING

Systems are usually designed to meet some type of specification—for example, position and velocity requirements, product composition, temperature, pressure and flow requirements—and in order to achieve this design, methods or techniques are employed from the particular branch of engineering involved; in our case this usually means electrical engineering, mechanical engineering and chemical engineering. In most cases only steady-state design methods are used. This results in a system that will produce the required specification in idealised conditions or circumstances only—that is, in theory only—because the design techniques are not exact, but are based usually on simplified empirical formulae; disturbances upon the system are neglected, as also are variations in the raw materials that are to be used in the process.

In practice, these factors would cause the actual product specification to be in error and in most cases this would lead to rejection by the customer. To make good these errors and to pull the product back into the required specification tolerance it is necessary to use some form of feedback control (which was outlined in the previous chapter); the feedback control measures the error between the desired performance and the actual performance, amplifies this error in some way and then uses the result to drive or force the system so as to reduce the error. It is seen that, by its very essence, the nature of the control system is concerned with affecting the dynamic and steady-state behaviour of the system. Once the need for control has been accepted, it then becomes necessary to design the control system.

There are many analytical techniques available for the design of control systems and, as will become evident later, all without exception require a knowledge and understanding of the behaviour of the system to be controlled. Since it is not usually possible to develop a control system in conjunction with the actual system as a result of safety, cost and time considerations—and in many cases because the system does not exist—its behaviour is usually deduced by considering the physics and chemistry of the system and writing down the relevant equations of motion. For example, it is usual to apply Newton's laws of motion in the case of mechanical systems, Kirchhoff's laws for electrical systems and mass and heat balances for chemical systems.

Mathematical Modelling

In control engineering, these equations of motion are collectively known as the *mathematical model* of the system (usually a set of differential equations); this system, which will later on be transformed into the system transfer function, represents an approximation to the actual dynamic and steady-state behaviour of the system. The extent of this approximation depends upon the complexity of the mathematical model and the purposes to which it is to be used. In particular, for the purposes of control-system design, this model should show how the system output is affected by changes in the system input because, as was stated earlier, the aim of the control system is to reduce the error in output by means of changes in the system input.

Often mathematical models are constructed not only for control-system-design work but for other purposes—for example, to obtain a greater understanding of the behaviour of the system or to investigate some aspect of the system design and possibly amend and improve it.

2.2 DESCRIPTION OF SIMPLE MECHANICAL SYSTEMS

Let us now consider how a mathematical model of a simple mechanical system—the spring–dashpot–mass arrangement shown in figure 2.1—is developed before we proceed to more complex modelling techniques.

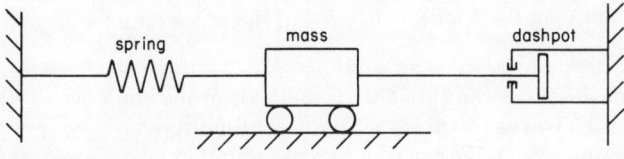

Figure 2.1

The spring–mass–dashpot system is chosen because most people have an understanding of the three components of the system whereas this understanding may not be readily available in the case of electrical-system components or chemical-system components. Let us consider the three components of figure 2.1 in turn.

The spring component is described by the stiffness of the spring, which is simply the relationship between the force applied to the spring and the amount of movement that results

$$\text{force } F \propto \text{movement } x \tag{2.1}$$

which, in the case of a linear spring, may be written as

$$F = Kx \text{ or } x = \frac{1}{K} F \tag{2.2}$$

where K is defined as the stiffness constant of the particular spring and is determined by the material, the size of the wire, the diameter and the spacing of the turns.

The second component, the dashpot, describes the effect experienced when we

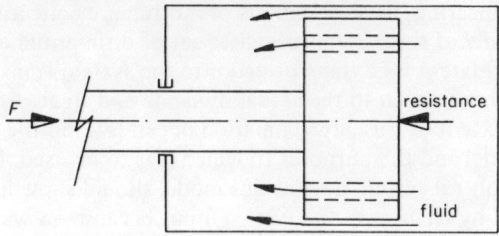

Figure 2.2

try to push an object through a thick fluid: the harder we push, the harder it becomes to move the object. In the idealised system shown in figure 2.2 the applied force F is pushing against a resistive force generated by the fluid pressing against the moving piston; this resistance is directly proportional to the velocity of the piston

$$F = Rv \tag{2.3}$$

where R is determined by the particular fluid in the dashpot, its size and shape. It is said that the dashpot exhibits viscous friction.

The final component, the mass, exhibits the property shown by mass itself when we try to move it: for example, a light sledge on smooth ice is relatively easy to accelerate whereas a heavy sledge needs a good deal more force. This relationship between the applied force F and the acceleration a may be written as

$$F = Ma \tag{2.4}$$

where the constant of proportionality is the mass of the component. This relationship is known as Newton's second law of motion.

The above three component properties are found in many important mechanical systems—for example, car suspension systems, vibration-absorption systems—though not necessarily in the idealised form considered here. For example, friction of some kind is nearly always present in real systems and even within our spring there will be friction. But it will be seen later that this idealisation of the real system is necessary if we are to obtain any kind of solution from our mathematical model; that is, the real world is far too complex to model exactly. In fact, it is in making realistic approximations to the behaviour of real systems so as to produce an idealised system that the greatness in engineering really lies.

Model of a Spring–Dashpot System

Let us now consider the two-component system consisting of a spring and a dashpot shown in figure 2.3; assume that the masses of the spring and dashpot can be ignored. The left-hand end of the spring is moved by an amount u and we wish to determine the resulting movement of the dashpot, x. The force exerted by the spring is

$$\begin{aligned} F_1 &= K(\text{extension}) \\ &= K(u - x) \end{aligned} \tag{2.5}$$

Mathematical Modelling

The force exerted by the dashpot is

$$F_2 = Rv$$
$$= R \frac{dx}{dt} \tag{2.6}$$

Since these must be equal, then

$$K(u - x) = R \frac{dx}{dt} \tag{2.7}$$

that is

$$R \frac{dx}{dt} + Kx = Ku \tag{2.8}$$

or

$$\mu \frac{dx}{dt} + x = u \tag{2.9}$$

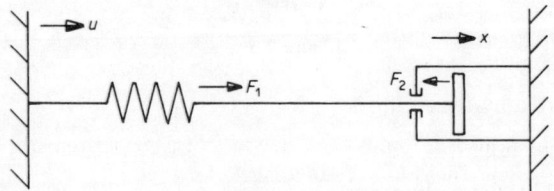

Figure 2.3

This will be recognised as a first-order linear differential equation and is in fact our mathematical model of the idealised two-component spring–dashpot system.

Given u, the model can be solved to obtain the dashpot movement $x(t)$, shown in figure 2.4 for various values of μ. This solution was obtained using a digital computer with a graphics terminal and figure 2.4 is in fact a copy of the traces as they appeared on the terminal. More will be said about the detailed solution of differential equations in the next chapter and use is made of the digital computer and graphics terminal throughout the book.

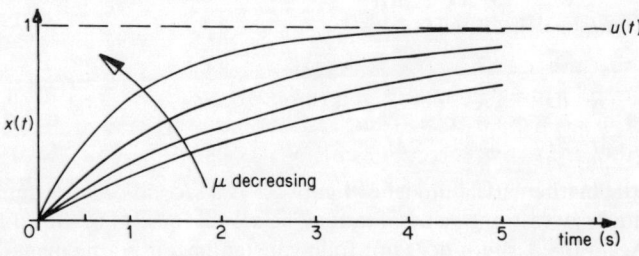

Figure 2.4

Thus from figure 2.4 it is seen that, following the movement u, the dashpot does not follow instantaneously but slowly moves to the position u after some lapse of time (due to the viscous-friction effects of the dashpot) determined by the value μ (= R/K). Later, this coefficient will be defined as the time constant of the system.

Model of a Spring-Dashpot-Mass System

If we now add a mass, the resulting system is as shown in figure 2.5 and, assuming that friction other than that in the dashpot can be neglected, we again wish to find the movement x of the dashpot following a movement u of the left-hand end of the spring.

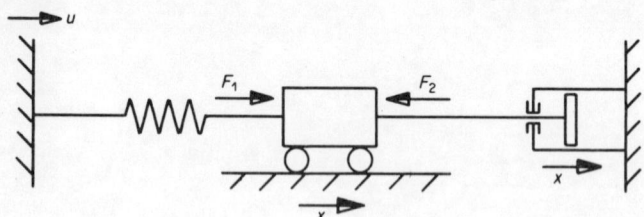

Figure 2.5

The force exerted by the spring is

$$F_1 = K(u - x) \tag{2.10}$$

The force exerted by the dashpot is

$$F_2 = Rv \tag{2.11}$$

Therefore by Newton's second law of motion

$$F_1 - F_2 = Ma \tag{2.12}$$

Hence

$$K(u - x) - Rv = Ma \tag{2.13}$$

that is

$$M \frac{d^2 x}{dt^2} + R \frac{dx}{dt} + Kx = Ku \tag{2.14}$$

or

$$\frac{d^2 x}{dt^2} + \frac{R}{M} \frac{dx}{dt} + \frac{K}{M} x = \frac{K}{M} u \tag{2.15}$$

In this case the mathematical model of figure 2.5 is a second-order differential equation, which, given some value of u, may be solved for $x(t)$ as shown in figure 2.6. Again the dashpot does not follow instantaneously; the main difference from the previous case is that because of the presence of the mass oscillatory behaviour can be obtained. More will be said about this later.

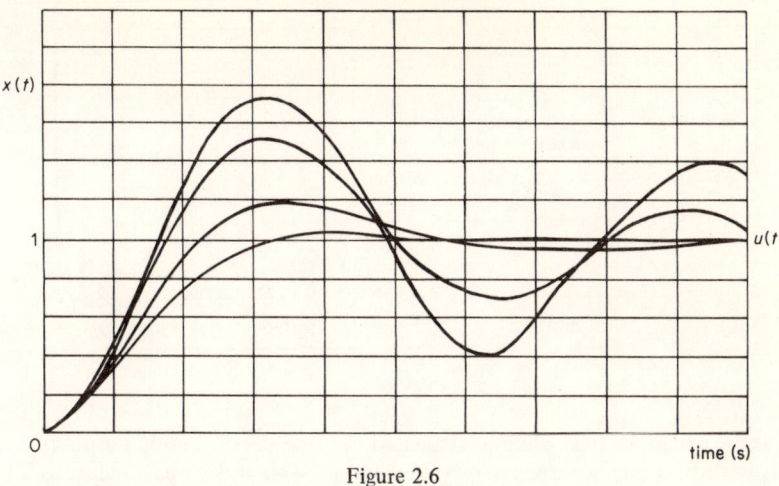
Figure 2.6

2.3 MORE ABOUT MATHEMATICAL MODELLING

In producing a mathematical model of a physical system certain fundamental steps are found to lead to models that show a satisfactory resemblance to the actual system but are very much simplified.

(1) A conceptual model is first specified, the building blocks of the model are idealisations of the events or phenomena occurring in the real system being modelled. These blocks could represent mass, spring, dashpot systems, electrical inductances and capacitances or the stirred tank of chemical engineering. This step results in a model essentially like the real system but much easier to solve mathematically.

(2) The relationships implied in the conceptual model are now expressed mathematically by applying the appropriate physical laws to each building block, thus obtaining the necessary differential equations of motion.

(3) These equations are expressed as a mathematical model of the system and the dynamic behaviour is obtained by solving the model. This behaviour is then compared with the actual or predicted system behaviour and, if necessary, adjustments are made to the model.

These steps lead to a mathematical model that resembles the salient features of the actual system, is simpler than the actual system, and therefore is more amenable to analytical study. This latter point is of profound importance since the main reason for the construction of mathematical models, as far as we are concerned, is their use as an aid to control-system design.

2.4 MODEL OF A TRAILER SUSPENSION SYSTEM

Before considering the modelling of electrical and chemical systems it will be beneficial to apply the above steps to an actual mechanical system that can be considered to **consist wholly of building blocks in the form of springs, dashpots**

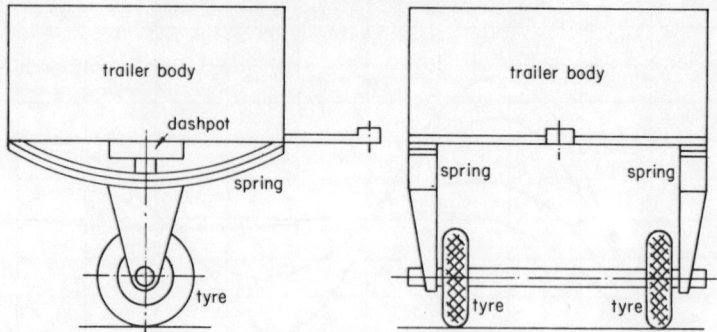

Figure 2.7

and masses, which we have already discussed in some detail. Let us suppose, then, that we wish to design a suspension system for a trailer and, in particular, to predict how the design would behave when experiencing bumpy road conditions. The trailer is shown in figure 2.7.

Firstly, a conceptual model of the system is developed, by assuming each tyre to behave as a linear spring and assuming that the mass of each tyre and associated suspension is lumped together into one mass M_2. The trailer body is considered as a mass M_1 that can move up and down and can also rotate, and the bumpy road is represented by position variations to each tyre separately. All this is depicted in figure 2.8.

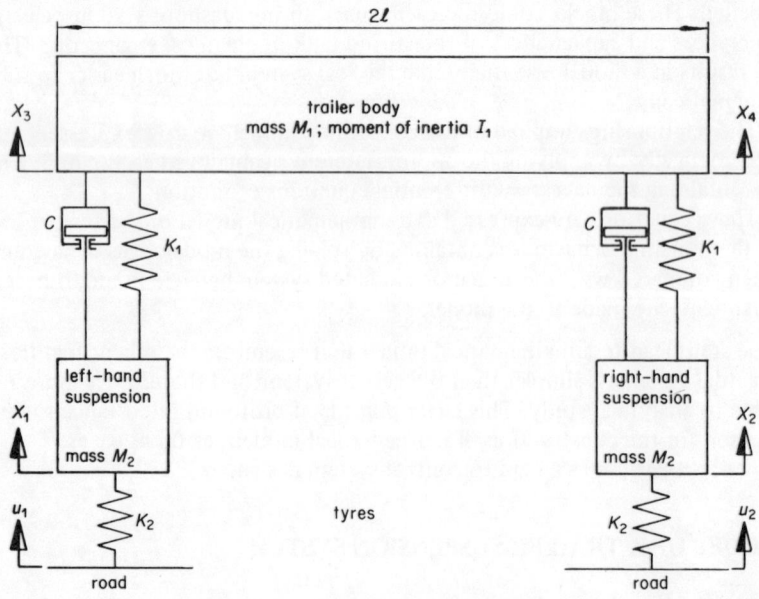

Figure 2.8

Mathematical Modelling

The next step in the procedure is to express mathematically the relationships implied in the conceptual model. Thus, using the notation of figure 2.8 and applying Newton's second law of motion to the left-hand suspension, we have

(mass M_2) (acceleration of M_2) = sum of applied forces acting on M_2 (2.16)

that is

$$M_2 \frac{d^2 X_1}{dt^2} = K_2(u_1 - X_1) - K_1(X_1 - X_3) - C\left(\frac{dX_1}{dt} - \frac{dX_3}{dt}\right) \quad (2.17)$$

and similarly for the right-hand suspension

$$M_2 \frac{d^2 X_2}{dt^2} = K_2(u_2 - X_2) - K_1(X_2 - X_4) - C\left(\frac{dX_2}{dt} - \frac{dX_4}{dt}\right) \quad (2.18)$$

The centre of gravity of the trailer body experiences both linear motion and rotational motion; if the left-hand end of the trailer moves an amount X_3 and the right-hand end an amount X_4 then the centre of gravity moves an amount

$$\frac{X_3 + X_4}{2}$$

vertically, and is rotated through an angle of approximately

$$\frac{X_3 - X_4}{l} \text{ rad}$$

in a clockwise direction.

Thus considering the vertical motion of the trailer-body centre of gravity

$$M_1 \frac{d^2}{dt^2}\left(\frac{X_3 + X_4}{2}\right) = K_1(X_1 - X_3) + K_1(X_2 - X_4)$$

$$+ C\left(\frac{dX_1}{dt} - \frac{dX_3}{dt}\right) + C\left(\frac{dX_2}{dt} - \frac{dX_4}{dt}\right) \quad (2.19)$$

For rotational motion, Newton's law states that

(moment of inertia of body) (angular acceleration) = sum of applied moments acting on the body (2.20)

that is

$$I \frac{d^2}{dt^2}\left(\frac{X_3 - X_4}{l}\right) = K_1 l(X_1 - X_3) - K_1 l(X_2 - X_4)$$

$$+ C\left(\frac{dX_1}{dt} - \frac{dX_3}{dt}\right) l - C\left(\frac{dX_2}{dt} - \frac{dX_4}{dt}\right) l \quad (2.21)$$

12 Introduction to Control Theory

The above four second-order differential equations, when collected together, constitute the mathematical model of the trailer suspension system; given values for the parameters K_1, K_2, C, l, M_1 and M_2 and the input movements u_1 and u_2, this model may be solved on a digital (or analogue) computer to give the resulting movements X_3 and X_4 and hence the motion of the trailer. If this resulting motion is not satisfactory then new parameter values may be chosen and the model solved again: in this way an acceptable design may be obtained. Figure 2.9 shows the system behaviour following an instantaneous change in u_1, which can be taken to represent the effect of the left-hand tyre hitting the kerb.

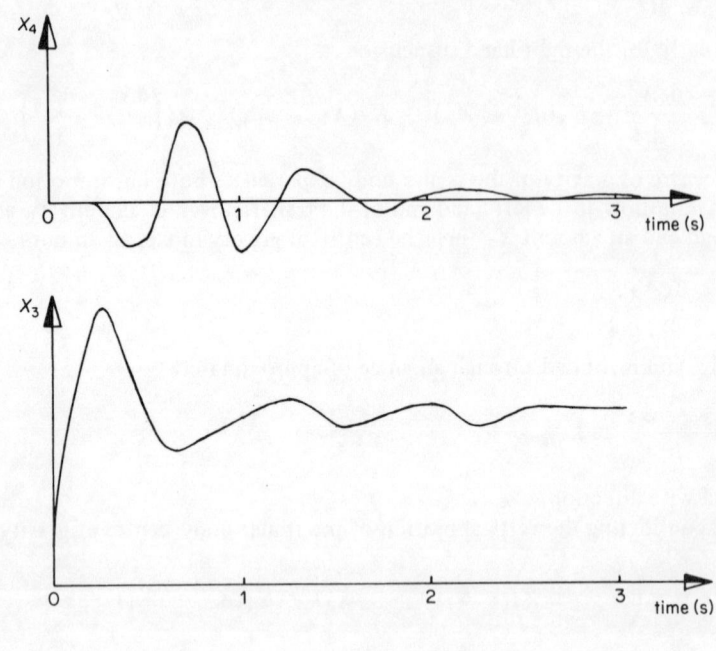

Figure 2.9

2.5 MODELLING OF SIMPLE ELECTRICAL SYSTEMS

Let us now consider how to produce a mathematical model of a simple electrical system consisting of ideal passive elements—a resistor, a capacitor and an inductor —assuming them all to be linear, as shown in figure 2.10.

When a voltage is applied to the two ends of a resistor, a current flows; the magnitude of this current is proportional to the applied field and the cross-sectional area of the conducting metal of the resistor, and is inversely proportional to its length. Thus

$$I = k \frac{vA}{l} \qquad (2.22)$$

Mathematical Modelling 13

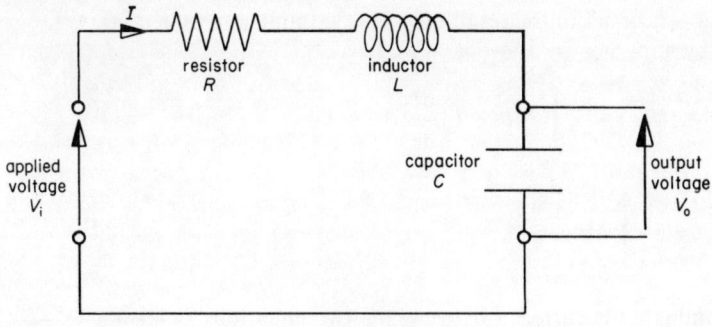

Figure 2.10

that is

$$v = RI \tag{2.23}$$

which is known as Ohm's Law; R is the resistance.

The voltage across a capacitor is proportional to the charge; the constant of proportionality depends upon the dimensions of the capacitor plates, their spacing and the dielectric material between them. For a given capacitor this constant is known as the capacitance C of the element and therefore

$$v = q/C \tag{2.24}$$

Since the total charge on the capacitor will be equal to the integral of the flow of current

$$v = \frac{1}{C} \int I \, dt \tag{2.25}$$

An inductor is constructed in the form of a coil and has the property that between its terminals there always exists a voltage proportional to the rate of change of current flowing through it. Thus

$$v \propto \frac{dI}{dt} \tag{2.26}$$

or

$$v = L \frac{dI}{dt} \tag{2.27}$$

where L is the inductance.

Before we can construct a mathematical model of the electrical system in figure 2.10 we need to be familiar with the equilibrium and compatibility relations that exist within electrical networks: these are found in Kirchhoff's Laws.

Kirchhoff's current law. When two or more conductors meet (at a node) the total current flowing into the node is equal to the total current flowing away from it.

Kirchhoff's voltage law. In passing around any complete circuit (or closed path) within the network, the algebraic sum of the voltage drops of the separate components is equal to the resultant e.m.f. within the circuit.

Thus by applying the above laws we have

$$V_i = IR + \frac{1}{C}\int I\,dt + L\frac{dI}{dt} \tag{2.28}$$

and

$$V_o = \frac{1}{C}\int I\,dt \tag{2.29}$$

Now eliminate the current I between the two equations by letting

$$I = C\frac{dV_o}{dt} \quad \text{and} \quad \frac{dI}{dt} = C\frac{d^2V_o}{dt^2} \tag{2.30}$$

to give

$$V_i = CR\frac{dV_o}{dt} + V_o + LC\frac{d^2V_o}{dt^2} \tag{2.31}$$

that is

$$\frac{d^2V_o}{dt^2} + \frac{R}{L}\frac{dV_o}{dt} + \frac{1}{LC}V_o = \frac{1}{LC}V_i \tag{2.32}$$

which is of exactly the same form as the mathematical model of the spring–dashpot–mass system derived earlier. Thus the behaviour of the output voltage V_o when there is a change in the input voltage V_i is as shown in figure 2.11, which is a replica of figure 2.6.

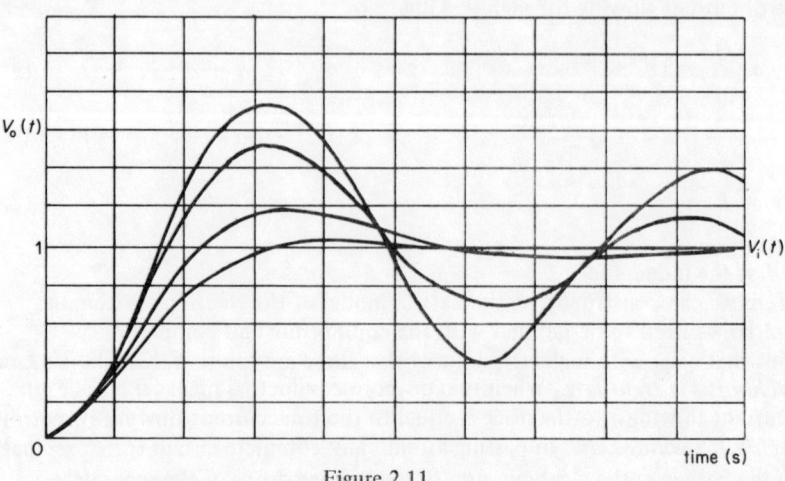

Figure 2.11

Mathematical Modelling

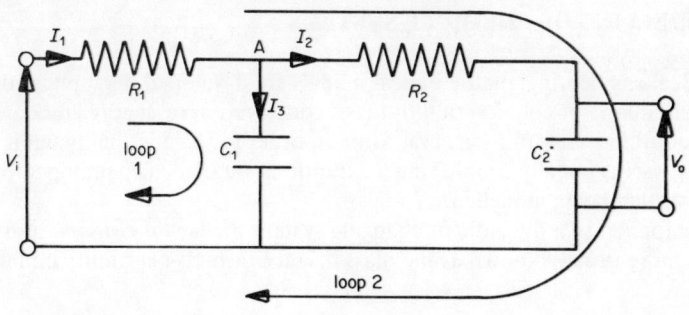

Figure 2.12

Another Electrical System

As a further example to show the use of Kirchhoff's laws let us consider the electrical system shown in figure 2.12, which consists of two RC circuits in series. Assuming that no current is drawn at the output and applying Kirchhoff's laws we have the following relations: at node A

$$I_1 = I_2 + I_3 \tag{2.33}$$

in loop 1

$$V_i = I_1 R_1 + \frac{1}{C_1} \int I_3 \, dt \tag{2.34}$$

and in loop 2

$$V_i = I_1 R_1 + I_2 R_2 + \frac{1}{C_2} \int I_2 \, dt \tag{2.35}$$

and

$$V_o = \frac{1}{C_2} \int I_2 \, dt \tag{2.36}$$

Eliminating I_1, I_2 and I_3 from the above equations and re-arranging gives

$$R_1 C_1 R_2 C_2 \frac{d^2 V_o}{dt^2} + (R_1 C_1 + R_2 C_2 + R_1 C_2) \frac{dV_o}{dt} + V_o = V_i \tag{2.37}$$

that is

$$\frac{d^2 V_o}{dt} + \frac{R_1 C_1 + R_2 C_2 + R_1 C_2}{R_1 C_1 R_2 C_2} \frac{dV_o}{dt} + \frac{1}{R_1 C_1 R_2 C_2} V_o$$
$$= \frac{1}{R_1 C_1 R_2 C_2} V_i \tag{2.38}$$

Thus for an electrical system consisting of two RC circuits in series the mathematical model relating output voltage to input voltage is a second-order differential equation.

2.6 MODELLING OF CHEMICAL SYSTEMS

In chemical engineering, mathematical models are developed by application of the fundamental laws of conservation of mass, conservation of energy and conservation of momentum in their most general form, in order to include the dynamic effects that are present. Usually simplifying assumptions are made depending upon the specific system being modelled.

When applied to a dynamic or changing system the law of conservation of mass, which is more usually known as the mass-balance or mass-continuity equation, states that

> mass flow into the system − mass flow out of the system
> = rate of change of mass accumulated within the system (2.39)

that is

$$W_i - W_o = \frac{d}{dt}(V\rho)$$

and for constant density

$$W_i - W_o = \rho \frac{d}{dt} V \qquad (2.40)$$

which, when steady-state conditions prevail, reduces simply to 'what goes in, must come out' since the right-hand side of the equation is zero under these conditions.

The law of conservation of energy or energy equation states that

> flow of energy into the system − flow of energy out of the system
> + heat added to the system − work done by the system
> = rate of change of energy accumulated within the system (2.41)

that is

$$W_i h_i - W_o h_o + Q - WD = \frac{d}{dt}(\rho V h)$$

and for constant density

$$W_i h_i - W_o h_o + Q - WD = \rho \frac{d}{dt}(Vh) \qquad (2.42)$$

where h is defined as the enthalpy. Often this general form reduces to what is known as an 'enthalpy balance'.

The final equation to be developed is known as the law of conservation of momentum or the momentum equation and is derived by application of Newton's second law of motion to fluid-flow systems. It states that

> the net sum of forces along a certain direction
> = the rate of change of momentum in the same direction (2.43)

Mathematical Modelling

that is

$$\sum_i F_i = \frac{d}{dt}(Mv) \tag{2.44}$$

where v is the velocity. In general the forces present are those due to pressure, gravity and friction and in many cases the momentum equation reduces to a friction equation relating pressure drop across the system to the mass flow through the system, that is

$$p_i - p_o = KW^2 \tag{2.45}$$

where K is a constant related to the friction factor of the system.

Let us now apply the above laws to the flow system shown in figure 2.13,

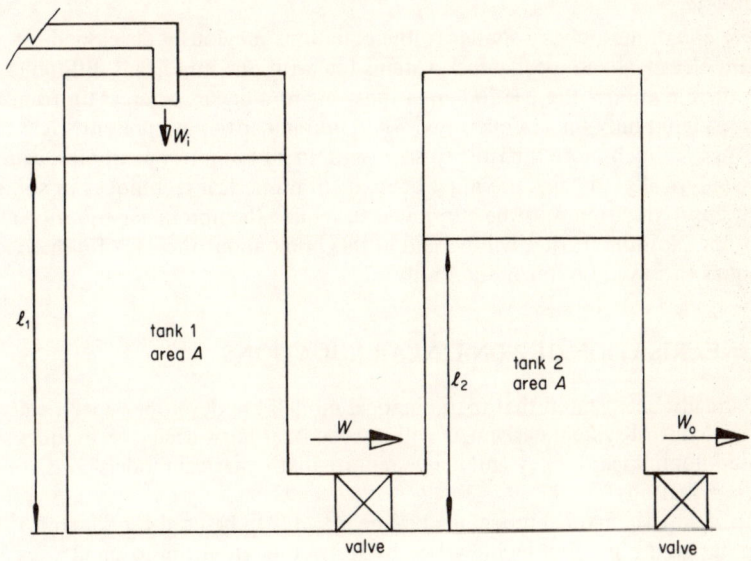

Figure 2.13

which consists of two tanks in series; the first experiences an inflow W_i and an outflow W, and the second an inflow W and an outflow W_o. There is no heat added to or lost from the system and so there is no need to consider an energy balance.

Application of the continuity equation gives: for tank 1

$$W_i - W = \rho \frac{dV_1}{dt} = \rho A \frac{dl_1}{dt} \tag{2.46}$$

and for tank 2

$$W - W_o = \rho A \frac{dl_2}{dt} \tag{2.47}$$

The flows W and W_o are obtained from a form of friction equation written for each valve (or restriction)

$$W = C\sqrt{(p_1 - p_2)} \tag{2.48}$$

and

$$W_o = C_o\sqrt{(p_2 - p_{atm})} \tag{2.49}$$

The pressures p_1 and p_2 are due to the heads of liquid in each tank, that is, they are hydrostatic pressures; thus

$$p_1 = \rho l_1; p_2 = \rho l_2 \tag{2.50}$$

The above equations constitute a mathematical model of the chemical system consisting of two tanks in series and may be solved to obtain the response of the outlet flow W_o and the levels in the tanks l_1 and l_2, following some change in the inlet flow W_i.

These equations are very similar to the equations previously developed to represent electrical and mechanical systems but with one important difference: two of the equations—the friction equations—are non-linear, whereas up to now we have considered only linear equations. When non-linearities are present, the equations are much more difficult to solve and, in all but a few examples, cannot be solved analytically; instead, use must be made of numerical techniques or some form of approximation must be employed to reduce the non-linear equations to linear ones. Nothing further will be said in this book about the use of numerical techniques to solve non-linear equations.

2.7 LINEARISATION OF NON-LINEAR EQUATIONS

It has already been stated that mathematical models are developed specifically for use in control-system design. As will be seen later most design techniques are based upon linear theory and hence require mathematical models consisting of linear equations.

Thus if a mathematical model is non-linear and is to be used for control-system design it must first be linearised by restricting attention to small perturbations of the mathematical model around a given reference state. This is done quite simply in the following manner. The non-linear functions are expanded into Taylor-series expansions around the reference state (or steady-state operating level) and then all terms after the first partial derivatives, that is, first-order terms, are neglected.

Consider a non-linear function $f(x_1, x_2)$ of the variables x_1 and x_2, which is to be linearised about the reference state $\bar{x}_1, \bar{x}_2$. First expand the function into a Taylor's expansion to give

$$f(x_1, x_2) = f(\bar{x}_1, \bar{x}_2) + \left(\frac{\partial f}{\partial x_1}\right)_{\bar{x}_1, \bar{x}_2} (x_1 - \bar{x}_1)$$
$$+ \left(\frac{\partial f}{\partial x_2}\right)_{\bar{x}_1, \bar{x}_2} (x_2 - \bar{x}_2) + \left(\frac{\partial^2 f}{\partial x_1^2}\right)_{\bar{x}_1, \bar{x}_2} \frac{(x_1 - \bar{x}_1)^2}{2!} + \ldots \tag{2.51}$$

where $(\partial f/\partial x_1)_{\bar{x}_1, \bar{x}_2}$ means $(\partial f/\partial x_1)$ evaluated at the point $x_1 = \bar{x}_1, x_2 = \bar{x}_2$ and is just a numerical quantity. Now neglect the terms of second and higher order to give

$$\left(\frac{\partial f}{\partial x_1}\right)_{\bar{x}_1, \bar{x}_2} (x_1 - \bar{x}_1) + \left(\frac{\partial f}{\partial x_2}\right)_{\bar{x}_1, \bar{x}_2} (x_2 - \bar{x}_2) = 0 \qquad (2.52)$$

that is

$$A\Delta x_1 + B\Delta x_2 = 0 \qquad (2.53)$$

where Δx_1 and Δx_2 are the perturbed variables with reference to $\bar{x}_1$ and $\bar{x}_2$, respectively, and

$$\left. \begin{array}{l} x_1 = \bar{x}_1 + \Delta x_1 \\ x_2 = \bar{x}_2 + \Delta x_2 \end{array} \right\} \qquad (2.54)$$

This linearisation is represented in figure 2.14 for the case of a function of one variable only.

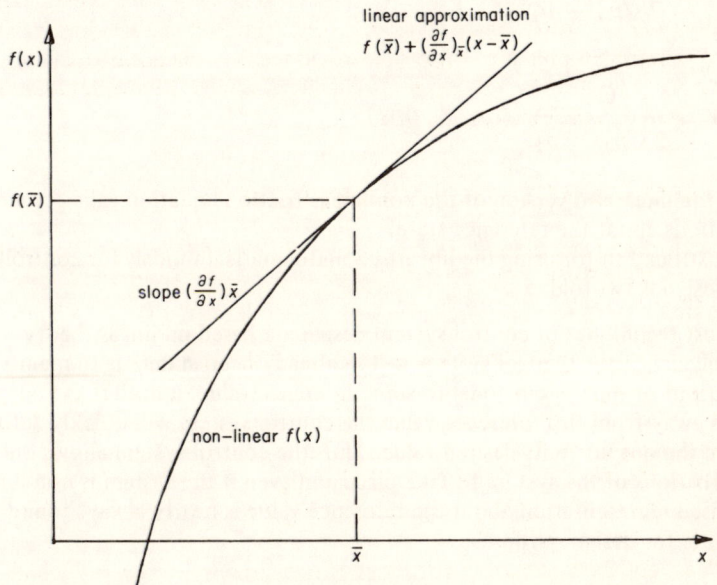

Figure 2.14

Let us now linearise the first of the non-linear friction equations

$$W = C\sqrt{(p_1 - p_2)} \qquad (2.55)$$

about the reference state $\bar{W}, \bar{p}_1, \bar{p}_2$. Let

$$f = W - C\sqrt{(p_1 - p_2)} \qquad (2.56)$$

Then

$$\left(\frac{\partial f}{\partial W}\right)_{\overline{W}, \overline{p}_1, \overline{p}_2} = 1 \tag{2.57}$$

$$\left(\frac{\partial f}{\partial p_1}\right)_{\overline{W}, \overline{p}_1, \overline{p}_2} = \left(-\frac{C}{2\sqrt{(p_1 - p_2)}}\right)_{\overline{W}, \overline{p}_1, \overline{p}_2}$$

$$= -\frac{C}{2\sqrt{(\overline{p}_1 - \overline{p}_2)}} \tag{2.58}$$

and

$$\left(\frac{\partial f}{\partial p_2}\right)_{\overline{W}, \overline{p}_1, \overline{p}_2} = \left(\frac{C}{2\sqrt{(p_1 - p_2)}}\right)_{\overline{W}, \overline{p}_1, \overline{p}_2} = \left(\frac{C}{2\sqrt{(\overline{p}_1 - \overline{p}_2)}}\right) \tag{2.59}$$

Therefore

$$\Delta W - \frac{C}{2\sqrt{\overline{p}_1 - \overline{p}_2}} (\Delta p_1 - \Delta p_2) = 0 \tag{2.60}$$

that is

$$\Delta W = \frac{C}{2\sqrt{\overline{p}_1 - \overline{p}_2}} (\Delta p_1 - \Delta p_2) \tag{2.61}$$

which is the linearised version of the non-linear friction equation valid for small perturbations about the reference state.

The justification for using the linearised mathematical models for control-system design is twofold.

(1) Most techniques of control-system design are based on linear theory.
(2) The aim of the control system, as has already been stated, is to maintain the output of the system equal to some reference value; indeed if the output moves away from this reference value the control system will quickly act to restore the output to its desired value. Thus the control system allows only small perturbations of the system to take place and even if the system is non-linear a linearised representation about the reference value is nearly always found to be adequate for design purposes.

2.8 CONCLUDING REMARKS

This chapter has been concerned with the development of mathematical models of mechanical, electrical and chemical systems, which are a necessary requirement in the design of control systems. In the next chapter these mathematical models are transformed into transfer functions relating the input and output of the system, which are then used extensively throughout the book.

Mathematical Modelling

REFERENCES

Cannon, R. H., *Dynamics of Physical Systems* (McGraw–Hill, New York, 1967).
Elgerd, O. I., *Control Systems Theory* (McGraw–Hill, New York, 1967).
Luyber, W. L., *Process Modelling, Simulation and Control for Chemical Engineers* (McGraw-Hill, New York, 1973).

PROBLEMS

2.1 Derive a complete set of equations (mathematical model) for the mechanical system in figure 2.15, which represents a vibration absorber. Assume idealised components throughout.

2.2 Figure 2.16 represents a conceptual model of a delicate instrument in which there is viscous friction present between the masses M_1 and M_2 and the frame M_3. Develop a mathematical model.

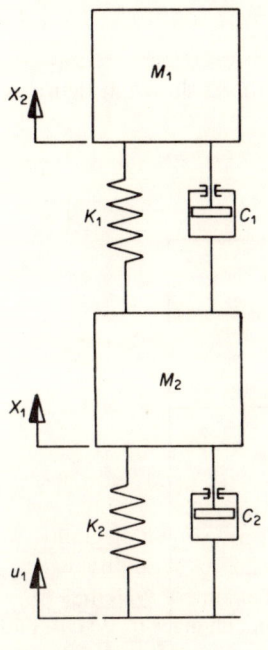

Figure 2.15

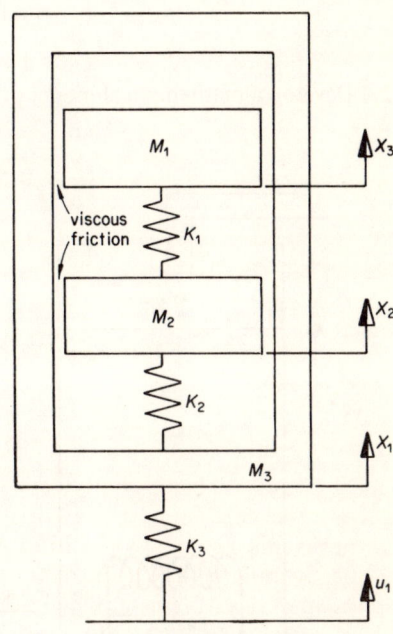

Figure 2.16

2.3 By applying Kirchhoff's laws develop a mathematical model for each of the networks shown in figure 2.17.

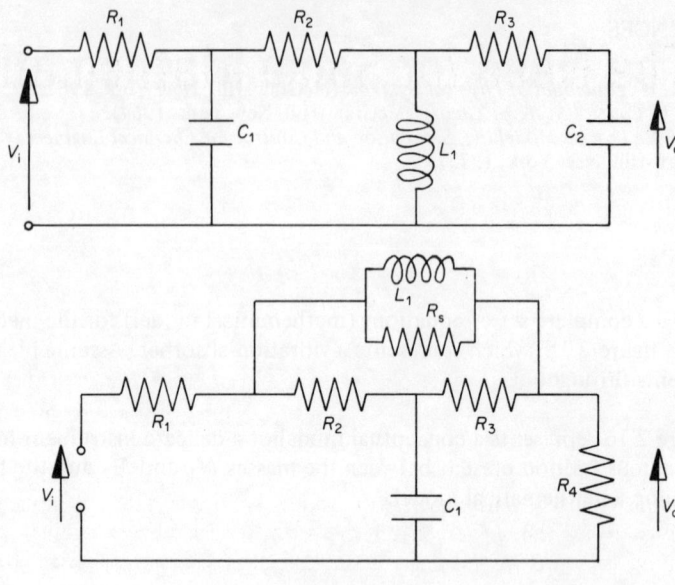

Figure 2.17

2.4 Develop a mathematical model of the flow-thermal process shown in figure 2.18.

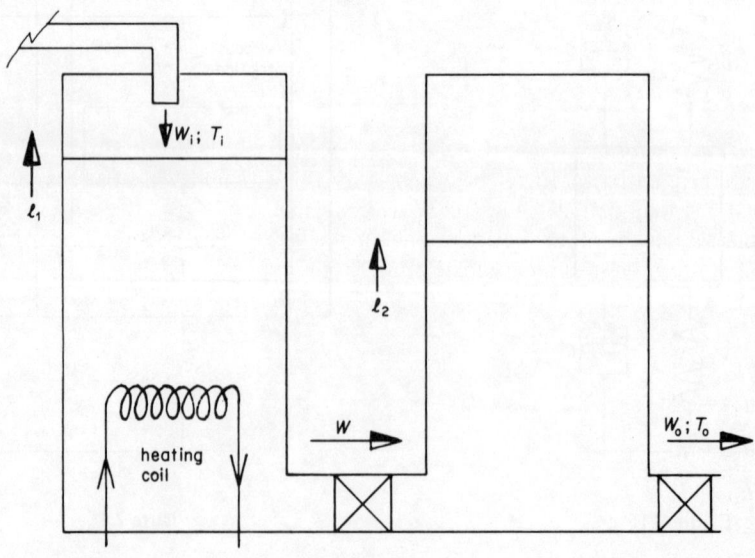

Figure 2.18

3 The Need for Some Mathematics

In the previous chapter, mathematical models of various physical systems (and reasons for obtaining them) were given; the models were either obtained directly in the form of first- and second-order linear differential equations with constant coefficients or were brought to that form by linearisation (or perturbation) methods. In general, mathematical models of complicated systems will take the form of nth order linear differential equations, which may be written as follows

$$a_0 \frac{d^n x}{dt^n} + a_1 \frac{d^{n-1} x}{dt^{n-1}} + a_2 \frac{d^{n-2} x}{dt^{n-2}} + \ldots + a_{n-1} \frac{dx}{dt} + a_n x$$
$$= b_0 \frac{d^m u}{dt^m} + b_1 \frac{d^{m-1} u}{dt^{m-1}} + \ldots + b_{m-1} \frac{du}{dt} + b_m u \qquad (3.1)$$

where $x(t)$ is the output from the system; $u(t)$ is the input to the system; d^r/dt^r is the rth differential coefficient with respect to time; n and m are indices ($n \geqslant m + 1$); the coefficients a_i and b_i are constant.

In the case of the mathematical model of the spring–dashpot–mass system in figure 2.5, it can be seen by comparison that

$n = 2, \quad m = 0$
$a_0 = M$
$a_1 = R$
$a_2 = K$
$b_0 = K$

that is

$$M \frac{d^2 x}{dt^2} + R \frac{dx}{dt} + Kx = Ku \qquad (3.2)$$

Introduction to Control Theory

It is the purpose of this chapter to show how solutions (for example, the time responses of figures 2.4 and 2.6) of equations such as 3.1 and 3.2 are obtained using both the so-called classical method and the method of Laplace transformation. Applications are considered and the chapter also includes methods of determining roots of polynomials and partial-fraction expansions. Some simple theory of algebraic equations is also considered.

3.1 CLASSICAL METHOD OF SOLUTION OF LINEAR DIFFERENTIAL EQUATIONS

The solution of a linear differential equation using the classical method has two components: the *complementary function* and the *particular integral*. The latter is obtained by solving equation 3.1 as it stands and represents the steady-state or forced solution of the equation whilst the former is the solution of the equation with the right-hand side made equal to zero. This equation is called the *characteristic equation* and its solution represents the transient behaviour of the system as it moves from its initial state to a final steady state. It is independent of the forcing input.

Consider first, by way of example and before obtaining a general solution of equation 3.1, the solution of the first-order differential equation that represents the behaviour of the spring–dashpot system in figure 2.3,

$$\mu \frac{dx}{dt} + x = u \tag{3.3}$$

with zero initial conditions, that is, $x(0) = 0$; in equation 3.3, u is a constant, and $\mu = R/K$.

The complementary function is the solution of the characteristic equation

$$\mu \frac{dx}{dt} + x = 0 \tag{3.4}$$

and it follows that x must have the same form as its derivative in order to obtain cancellation; that is, assume

$$x_c = Ae^{mt} \tag{3.5}$$

Therefore

$$\mu A m e^{mt} + A e^{mt} = 0 \tag{3.6}$$

Therefore

$$\mu m + 1 = 0 \tag{3.7}$$

that is

$$m = -\frac{1}{\mu}$$

and
$$x_c = Ae^{-(1/\mu)t} \tag{3.8}$$

Equation 3.7 is known as the *auxiliary equation* and equation 3.8 represents the complementary function to within an arbitrary multiplying constant A (which is determined later).

The particular integral is the solution of equation 3.3, and in general is obtained using experience, but in this case, since the forcing input is constant, choose

$$x_p = Bu \tag{3.9}$$

Therefore
$$0 + Bu = u$$
and
$$B = 1 \tag{3.10}$$

Thus the full solution is given by
$$\begin{aligned}x &= x_c + x_p \\ &= Ae^{-(1/\mu)t} + u\end{aligned} \tag{3.11}$$

and since, at time $t = 0$, $x(0) = 0$, then
$$0 = A + u$$
Therefore
$$A = -u$$
and it is seen that the arbitrary multiplying factor is obtained using knowledge of the initial conditions. Therefore

$$x(t) = (1 - e^{-(1/\mu)t})u \tag{3.12}$$

This relation is shown in figure 3.1 for various values of u.

The speed of response of the first-order system increases as μ decreases, that is, as R (damping) decreases and K (spring constant) increases, which is correct. In fact, μ is known as the *time constant* of the system and is equal to the time intersect of the tangent at the origin and the steady-state value, as shown in figure 3.1.

Consider now the general equation 3.1 but with the input function remaining constant, that is

$$a_0 \frac{d^n x}{dt^n} + a_1 \frac{d^{n-1} n}{dt^{n-1}} + a_2 \frac{d^{n-2} x}{dt^{n-2}} + \ldots + a_{n-1} \frac{dx}{dt}$$
$$+ a_n x = b_0 u \tag{3.13}$$

and as before (and for the same reason) choose $x_c = Ae^{mt}$. Substituting this value into the characteristic equation associated with equation 3.13 gives

$$Ae^{mt}(a_0m^n + a_1m^{n-1} + \ldots + a_{n-1}m + a_n) = 0 \qquad (3.14)$$

that is

$$a_0m^n + a_1m^{n-1} + \ldots + a_{n-1}m + a_n = 0 \qquad (3.15)$$

This is the nth-order auxiliary equation and gives rise to the n solutions or roots, $m_1, m_2, \ldots, m_n$. These roots are in general complex but, since all the coefficients of equation 3.13 are real, they must occur in conjugate pairs, that is, $\sigma \pm j\omega$.

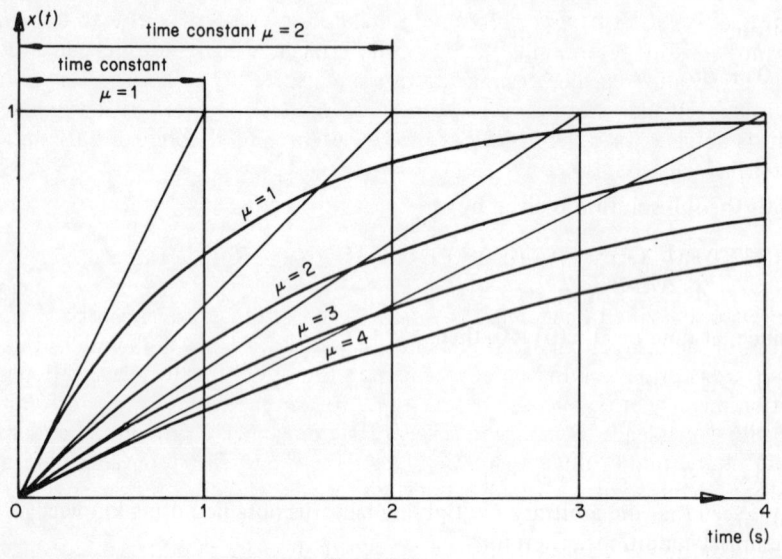

Figure 3.1

Thus the complementary function is

$$x_c = A_1 e^{m_1 t} + A_2 e^{m_2 t} + \ldots + A_n e^{m_n t} = \sum_{i=1}^{n} A_i e^{m_i t} \qquad (3.16)$$

a summation of the n solutions.

The particular integral is similar to the preceding case, that is

$$x_p = \frac{b_0}{a_n} u \qquad (3.17)$$

and therefore the full solution is

$$x = \sum_{i=1}^{n} A_i e^{m_i t} + \frac{b_0}{a_n} u \qquad (3.18)$$

and the n unknown coefficients $A_1, A_2, \ldots, A_n$ are obtained from knowledge of the n initial conditions, that is, $x(0)$, $dx(0)/dt$, $d^2x(0)/dt^2, \ldots, d^{n-1}x(0)/dt^{n-1}$.

The full solution as given by equation 3.18 will be stable, that is, it will remain

The Need for Some Mathematics

constant as time increases, if the exponential terms $e^{m_i t}$ decay. Therefore, assuming that the roots m_i are real, stability will be guaranteed if the m_i are negative, that is $m_i < 0$. For a complex root, that is, $m_i = \sigma_i + j\omega_i$, the solution will be stable if the real part is negative, that is, $\sigma_i < 0$. To summarise, the system that is represented by equation 3.13 will be stable if the real parts of the n roots of the auxiliary equation are negative. More will be said about stability later.

Comment on the Classical Method

There are two main disadvantages of proceeding with this classical approach. Firstly it is not always straightforward to obtain the n unknown coefficients of equation 3.18 and secondly determining the particular integral is difficult except in a few simple cases, and even then complications may arise.

It is also felt that the approach is too 'mathematical' and very little engineering insight is obtained into the behaviour of the system that is represented by the differential equation.

3.2 INTRODUCTION TO THE LAPLACE TRANSFORMATION

A more convenient and, in some cases, a more engineering-type approach to the solution of the differential equations under consideration lies in the use of the Laplace transformation. Instead of obtaining the solution of the initial differential equation directly, it is generally more straightforward to transform the differential equation, which leads to an algebraic equation rather than a differential one, to obtain the solution of this transformed equation, and finally to obtain the inverse transform of this solution, which is the solution of the original differential equation.

The operation of Laplace transformation is defined by the integral

$$\mathcal{L}[f(t)] = \int_0^\infty f(t)e^{-st} dt = F(s) \tag{3.19}$$

where $f(t)$ is a function of time and is equal to zero for $t < 0$; s is a complex variable; $F(s)$ is the function of s that is obtained on transforming $f(t)$; and $\mathcal{L}$ is the symbol to denote that $f(t)$ is to be Laplace transformed. The limits of integration in equation 3.19 are approached by a limiting process and hence, strictly speaking, the lower limit should read 0+, indicating that time t starts just after time equals zero. It should be noted that, in order for $F(s)$ to exist, $f(t)e^{-st}$ should approach zero as t approaches infinity. Using equation 3.19 it is possible to compile a table of Laplace-transform pairs, which will obviously avoid tedious repetition of the use of the definition. Such a table is given in the appendix (p. 226). The Laplace transform possesses a unique inverse transform given by

$$\mathcal{L}^{-1}[F(s)] = \frac{1}{2\pi j} \int_{\sigma-j\omega}^{\sigma+j\omega} f(s)e^{st} ds = f(t) \tag{3.20}$$

but normally, instead of using equation 3.20 to obtain the inverse transform, use is made of the table of transform pairs, reading from right to left rather than left to right.

To illustrate the use of this table, it can be seen that if

$$f(t) = e^{-at}, \quad F(s) = \frac{1}{s + a} \tag{3.21}$$

using the table directly, and if

$$F(s) = \frac{w}{s^2 + w^2}, \quad f(t) = \sin wt \tag{3.22}$$

using the table in an inverse manner. These relationships will be proved later.

3.3 SOME PROPERTIES OF THE LAPLACE TRANSFORMATION

Linearity and Superposition

If the Laplace transforms of any two functions $f_1(t)$ and $f_2(t)$ are $F_1(s)$ and $F_2(s)$, respectively, then

$$\begin{aligned}\mathcal{L}\left[a_1 f_1(t) + a_2 f_2(t)\right] &= a_1 \mathcal{L}\left[f_1(t)\right] + a_2 \mathcal{L}\left[f_2(t)\right] \\ &= a_1 F_1(s) + a_2 F_2(s)\end{aligned} \tag{3.23}$$

where a_1 and a_2 are constants. The proof follows directly from application of the definition of Laplace transformation, equation 3.19.

Differentiation

If the Laplace transform of the function $f(t)$ is $F(s)$, then

$$\mathcal{L}\left[\frac{df(t)}{dt}\right] = sF(s) - f(t)\Big|_{t=0} \tag{3.24}$$

where $f(t)_{t=0}$ means the value of the function $f(t)$ at $t = 0$ and is the initial value or condition of the function.

The proof follows by applying the rule for integration by parts

$$\int_0^\infty u \, dv = [uv]_0^\infty - \int_0^\infty v \, du \tag{3.25}$$

with $u = e^{-st}$ and $dv = [df(t)/dt] \, dt$. Thus

$$\mathcal{L}\left[\frac{df(t)}{dt}\right] = \int_0^\infty \frac{df(t)}{dt} e^{-st} \, dt = \left[e^{-st} f(t)\right]_0^\infty + \int_0^\infty f(t) s e^{-st} \, dt \tag{3.26}$$

$$= -f(t)\Big|_{t=0} + sF(s) \tag{3.27}$$

The Need for Some Mathematics

Similarly it can be shown that

$$\mathcal{L}\left[\frac{d^2 f(t)}{dt^2}\right] = s^2 F(s) - sf(t)\Big|_{t=0} - \frac{df(t)}{dt}\Big|_{t=0} \quad (3.28)$$

and, more generally, that

$$\mathcal{L}\left[\frac{d^k f(t)}{dt^k}\right] = s^k F(s) - s^{k-1} f(t)\Big|_{t=0} - s^{k-2} \frac{df(t)}{dt}\Big|_{t=0} - \cdots$$

$$\cdots - s \frac{d^{k-2} f(t)}{dt^{k-2}}\Big|_{t=0} - \frac{d^{k-1} f(t)}{dt^{k-1}}\Big|_{t=0} \quad (3.29)$$

Equations 3.27–3.29 are used extensively in the solution of linear differential equations.

Integration

If the Laplace transform of $f(t)$ is $F(s)$ then

$$\mathcal{L}\left[\int_0^t f(\tau)\,d\tau\right] = \frac{1}{s} F(s) \quad (3.30)$$

The proof follows by letting $u = \int_0^t f(\tau)d\tau$ and $dv = e^{-st}\,dt$ in equation 3.25; then

$$\mathcal{L}\left[\int_0^t f(\tau)d\tau\right] = \int_0^\infty \left[\int_0^t f(\tau)d\tau\right] e^{-st}\,dt$$

$$= \left[-\frac{1}{s} e^{-st} \int_0^t f(\tau)d\tau\right]_0^\infty + \int_0^\infty \frac{1}{s} e^{-st} f(t)\,dt \quad (3.31)$$

$$= \frac{1}{s} F(s)$$

since the first term on the right-hand side of equation 3.31 is zero for both $t = \infty$ and $t = 0$.

Time Translation

If the Laplace transform of $f(t)$ is $F(s)$, then

$$\mathcal{L}\left[f(t - \tau)\right] = e^{-s\tau} F(s) \quad (3.32)$$

provided that $f(t - \tau) = 0$ for $0 < t < \tau$, which is shown in figure 3.2. The proof follows from

$$\mathcal{L}\left[f(t - \tau)\right] = \int_0^\infty f(t - \tau)e^{-st}\,dt = \int_\tau^\infty f(t - \tau)e^{-st}\,dt \quad (3.33)$$

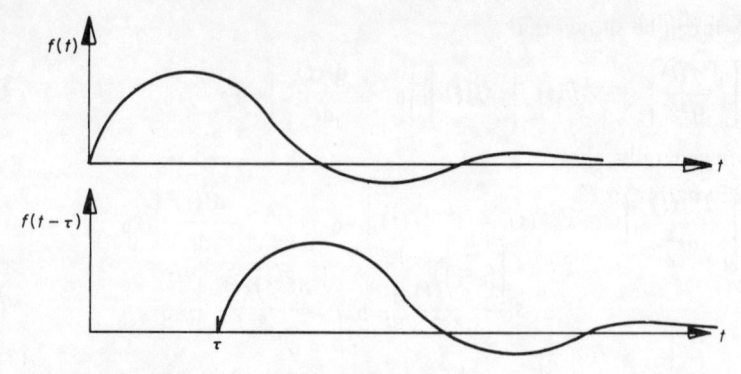

Figure 3.2

Now let $u = t - \tau$; then

$$\mathcal{L}[f(t - \tau)] \int_0^\infty f(u)e^{-s(u+\tau)}du = e^{-s\tau}\int_0^\infty f(u)e^{-su}du \qquad (3.34)$$
$$= e^{-s\tau}F(s)$$

Complex Translation

If the Laplace transform of $f(t)$ is $F(s)$ then

$$\mathcal{L}[e^{-at}f(t)] = F(s + a) \qquad (3.35)$$

The proof is left as an exercise for the reader.

Initial-value and Final-value Theorems

The initial-value theorem states that

$$\lim_{t \to 0} f(t) = \lim_{s \to \infty} sF(s) \qquad (3.36)$$

and the final-value theorem that

$$\lim_{t \to \infty} f(t) = \lim_{s \to 0} sF(s) \qquad (3.37)$$

provided that the limits exist in each of the cases. These theorems allow the time function at both $t = 0$ and $t = \infty$ to be obtained directly from the transform and can prove extremely useful.

3.4 LAPLACE TRANSFORMATIONS OF SOME TIME FUNCTIONS

The Impulse Function $\delta(t)$

$$\mathcal{L}[\delta(t)] = \int_0^\infty \delta(t)e^{-st}dt = \int_0^\infty \delta(t)dt = 1 \qquad (3.38)$$

since $\delta(t) = 0$ for all values of t except $t = 0$.

The Need for Some Mathematics

e Step Function a

$$\mathcal{L}[a] = \int_0^\infty ae^{-st}\,dt = \left[-\frac{a}{s}e^{-st}\right]_0^\infty = \frac{a}{s} \tag{3.39}$$

e Ramp Function at

$$\mathcal{L}[at] = \int_0^\infty ate^{-st}\,dt = \left[a\frac{t}{s}e^{-st}\right]_0^\infty + \int_0^\infty a\frac{1}{s}e^{-st}\,dt$$

$$= \left[-\frac{a}{s^2}e^{-st}\right]_0^\infty = \frac{a}{s^2} \tag{3.40}$$

e Exponential Function e^{-at}

$$\mathcal{L}[e^{-at}] = \int_0^\infty e^{-at}e^{-st}\,dt = \int_0^\infty e^{-(s+a)t}\,dt$$

$$= \left[-\frac{1}{s+a}e^{-(s+a)t}\right]_0^\infty = \frac{1}{s+a} \tag{3.41}$$

e Sine Function $\sin \omega t$

$$\mathcal{L}[\sin \omega t] = \int_0^\infty \sin \omega t\, e^{-st}\,dt$$

$$= \frac{1}{2j}\int_0^\infty [e^{j\omega t} - e^{-j\omega t}]e^{-st}\,dt$$

$$= \frac{1}{2j}\int_0^\infty [e^{-(s-j\omega)t} - e^{-(s+j\omega)t}]\,dt$$

$$= \frac{1}{2j}\left[\frac{1}{s-j\omega} - \frac{1}{s+j\omega}\right] = \frac{\omega}{s^2+\omega^2} \tag{3.42}$$

e Exponentially Decaying Sine Function $e^{-at}\sin \omega t$

ɔm the previous result

$$\sin \omega t = \frac{\omega}{s^2+\omega^2}$$

l using equation 3.35

$$\mathcal{L}[e^{-at}\sin \omega t] = \frac{\omega}{(s+a)^2+\omega^2} \tag{3.43}$$

The reader should verify this result by deriving it from first principles using the definition of the Laplace transform in a manner similar to the derivation of equation 3.42.

Before embarking upon the use of the Laplace transformation for the solution of linear differential equations, it is necessary to consider methods of determining roots of polynomials and partial-fraction expansions. Some simple theory of algebraic equations is also included. Readers familiar with these topics will be able to omit the next two sections without any loss of continuity but it is advised that they attempt problem 3.2 before doing so.

3.5 ROOTS OF POLYNOMIALS

In many branches of control theory it is necessary to be able to determine the roots of polynomials, that is, it is necessary to be able to factorise the polynomial. Later, we shall be considering in detail a graphical approach known as the root-locus method and also the use of digital computers. At the moment, though, we examine some numerical procedures that are considered useful for hand solution of polynomials up to, say, order five.

Some Simple Theory of Algebraic Equations

It is well known that any algebraic equation of the nth degree in s must have n roots. Let $s_1, s_2, \ldots, s_n$ be the n roots of the equation

$$a_0 s^n + a_1 s^{n-1} + a_2 s^{n-2} + \ldots + a_{n-1} + a_n = 0 \tag{3.44}$$

Then this equation must be identical with

$$a_0 (s - s_1)(s - s_2)(s - s_3) \ldots (s - s_{n-1})(s - s_n) = 0 \tag{3.45}$$

and multiplying out gives

$$a_0 \left(s^n - s^{n-1} \sum_{i=1}^{n} s_i + s^{n-2} \sum_{\substack{i,j=1 \\ i \neq j}}^{n} s_i s_j - \ldots \right) = 0 \tag{3.46}$$

Now comparing the coefficients of equations 3.44 and 3.46 gives

$$\sum_{i=1}^{n} s_i = -\frac{a_1}{a_0} \tag{3.47}$$

that is the sum of the roots is $-a_1/a_0$ and

$$\sum_{\substack{i,j=1 \\ i \neq j}}^{n} s_i s_j = \frac{a_2}{a_0} \tag{3.48}$$

that is, the sum of the roots taken two at a time is a_2/a_0, and so on until

$$\prod_{i=1}^{n} s_i = (-1)^n \frac{a_n}{a_0} \tag{3.49}$$

The Need for Some Mathematics   33

that is, the product of the roots is $(-1)^n a_n/a_0$. These results can be useful in establishing the approximate location of some or all of the roots of a polynomial.

Polynomials of Odd Order

When determining the roots of odd-order polynomials it is recognised, since all the coefficients are real, that the polynomial must have at least one real root, the remainder being real and/or complex, with the complex roots occurring as conjugate pairs. Thus, the real root is first determined using Newton's method and then factored out to leave a polynomial of even order.

Example 3.1
Determine the roots of the polynomial

$$f(s) = s^3 + 4.5s^2 + 13.5s + 35 = 0 \tag{3.50}$$

Initially the approximate location of the real root is established by determining a change of sign in the polynomial. Also, in this particular example, since all the coefficients are positive, this root must be negative. Let s_1 be the unknown root. Try various values of s_1: for $s_1 = 0$

$$f(s_1) = 35$$

for $s_1 = -1$

$$f(s_1) = -1 + 4.5 - 13.5 + 35 = 25$$

for $s_1 = -2$

$$f(s_1) = -8 + 18 - 27 + 35 = 18$$

for $s_1 = -3$

$$f(s_1) = -27 + 40.5 - 40.5 + 35 = 8$$

for $s_1 = -4$

$$f(s_1) = -64 + 72 - 54 + 35 = -11$$

Thus, the real root lies between -3 and -4.

Newton's method states that, given a first estimate of a root s^1, a second estimate s^2 may be obtained from the formula

$$s^2 = s^1 - \frac{f(s^1)}{f'(s^1)} \tag{3.51}$$

where $f'(s^1) = [\partial f(s)/\partial s]_{s=s^1}$ and subsequent estimates from

$$s^{i+1} = s^i - \frac{f(s^i)}{f'(s^i)} \tag{3.52}$$

In this way an accurate determination of the root may be obtained.
From equation 3.50

$$f(s) = s^3 + 4.5s^2 + 13.5s + 35$$
$$f'(s) = 3s^2 + 9s + 13.5$$

and let the first estimate be $s^1 = -3.4$. Then

$$f(-3.4) = -39.3 + 52 - 45.9 + 35 = 1.8$$
$$f'(-3.4) = 34.7 - 30.6 + 13.5 = 17.6$$

and therefore

$$s^2 = -3.4 - \frac{1.8}{17.6} = -3.5$$

Then $f(-3.5) = 0$ and therefore the unknown real root is $s_1 = -3.5$. This root is now divided out from the polynomial, as follows

```
                    s² +   s + 10
        ─────────────────────────────
s + 3.5 ) s³ + 4.5s² + 13.5s + 35
          s³ + 3.5s²
          ──────────
                s² + 13.5s
                s² +  3.5s
                ──────────
                     10s + 35
                     10s + 35
```

Therefore

$$(s^3 + 4.5s^2 + 13.5s + 35) = (s + 3.5)(s^2 + s + 10)$$

and the remaining roots are obtained directly from the quadratic formula as $s_2, s_3 = -\frac{1}{2} \pm j\sqrt{(39/4)}$. As a check, note that the sum of the roots is $(-3.5) + [-\frac{1}{2} + j\sqrt{(39/4)}] + [-\frac{1}{2} - j\sqrt{(39/4)}] = -4.5$ which agrees with equation 3.47.

Polynomials of Even Order

In this case, since all the roots may be complex, quadratic factors are obtained using Lin's method and then the quadratic formula may be used to obtain the actual roots.

Example 3.2
Determine the roots of the polynomial

$$s^4 + 2s^3 + 6s^2 + 5s + 6 = 0$$

In Lin's method, the first estimate of a quadratic factor is obtained from the last three terms of the polynomial being factored, that is

$$6s^2 + 5s + 6 \quad \text{or} \quad s^2 + \frac{5}{6}s + 1$$

and is divided out of the polynomial

The Need for Some Mathematics

$$
\begin{array}{r}
s^2 + 1.17s + 4.029 \\
s^2 + 0.83s + 1 \overline{\smash{\big)}\, s^4 + 2s^3 + 6s^2 + 5s + 6}\\
s^4 + 0.83s^3 + s^2 \\
\hline
1.17s^3 + 5s^2 + 5s \\
1.17s^3 + 0.971s^2 + 1.17s \\
\hline
4.029s^2 + 3.83s + 6 \quad \text{second estimate}\\
4.029s^2 + 3.344s + 4.029\\
\hline
\text{remainder}
\end{array}
$$

If the remainder is zero then the estimate is, in fact, a quadratic factor; if not, as in this case, a second estimate, which will be a closer approximation to the actual quadratic factor, is

$4.029s^2 + 3.83s + 6 \quad \text{or} \quad s^2 + 0.95s + 1.49$

Hence repeating the process

$$
\begin{array}{r}
s^2 + 1.05s + 3.51 \\
s^2 + 0.95s + 1.49 \overline{\smash{\big)}\, s^4 + 2s^3 + 6s^2 + 5s + 6}\\
s^4 + 0.95s^3 + 1.49s^2 \\
\hline
1.05s^3 + 4.51s^2 + 5s \\
1.05s^3 + s^2 + 1.56s \\
\hline
3.51s^2 + 3.44s + 6 \quad \text{third estimate}\\
3.51s^2 + 3.33s + 5.23\\
\hline
\text{remainder}
\end{array}
$$

Similarly the third estimate is

$3.51s^2 + 3.44s + 6 \quad \text{or} \quad s^2 + 0.98s + 1.71$

The process is continued until the remainder is small or zero. The quadratic factors are

$(s^2 + s + 2)(s^2 + s + 3)$

3.6 PARTIAL-FRACTION EXPANSIONS

When a ratio $f(s)/g(s)$, where $f(s)$ and $g(s)$ are polynomials and the order of $g(s)$ is *greater* than the order of $f(s)$, is expressed as a sum of two or more simpler ratios according to certain rules, the ratio is said to be resolved into *partial fractions*. Knowledge of partial fractions is required when determining inverse

Laplace transforms. Let

$$\frac{f(s)}{g(s)} = \frac{b_0 s^m + b_1 s^{m-1} + \ldots + b_{m-1} s + b_m}{s^n + a_1 s^{n-1} + \ldots + a_{n-1} s + a_n} \quad (n > m) \tag{3.53}$$

$$= \frac{b_0 s^m + b_1 s^{m-1} + \ldots + b_{m-1} s + b_m}{(s + s_1)(s + s_2)^k (s^2 + as + b)(s^2 + cs + d)^l}$$

then the partial-fraction expansion will be of the form

$$\frac{f(s)}{g(s)} = \frac{A_1}{(s + s_1)} + \frac{B_1}{(s + s_2)} + \frac{B_2}{(s + s_2)^2} + \ldots + \frac{B_k}{(s + s_2)^k}$$

$$+ \frac{C_1 s + D_1}{s^2 + as + b} + \frac{E_1 s + F_1}{s^2 + cs + d} + \frac{E_2 s + F_2}{(s^2 + cs + d)^2} + \ldots$$

$$+ \frac{E_l s + F_l}{(s^2 + cs + d)^l} \tag{3.54}$$

where $A_1, B_1, \ldots, B_k, C_1, D_1, E_1, F_1, \ldots, E_l, \ldots, F_l$ are constants whose values have to be determined. There are two approaches to this problem. The basis for the first is that equations 3.53 and 3.54 are identical and hence by giving suitable values to s or by equating coefficients the constants may be determined. The second approach is concerned with calculating the *residue* of each factor (or as it is more loosely termed, applying the 'cover-up' rule) and in many cases is less complicated than the first. Usually, however, a combination of the two approaches is employed.

Method 1

This is illustrated by the following example.

Example 3.3
Express the following ratio in partial fractions

$$\frac{s^2 + 2s - 5}{s(s + 1)(s + 5)^2}$$

Let

$$\frac{s^2 + 2s - 5}{s(s + 1)(s + 5)^2} \equiv \frac{A}{s} + \frac{B}{s + 1} + \frac{C}{s + 5} + \frac{D}{(s + 5)^2}$$

$$\equiv \frac{A(s + 1)(s + 5)^2 + Bs(s + 5)^2 + Cs(s + 1)(s + 5) + Ds(s + 1)}{s(s + 1)(s + 5)^2}$$

Therefore

$$s^2 + 2s - 5 \equiv A(s + 1)(s + 5)^2 + Bs(s + 5)^2 + Cs(s + 1)(s + 5) + Ds(s + 1)$$

Since this is an identity it is true for all values of s; thus, when $s = 0$

$$-5 \equiv A(1)(25)$$

and therefore $A = -1/5$; when $s = -1$

$$-6 \equiv B(-1)(16)$$

and therefore $B = 3/8$; when $s = -5$

$$10 \equiv D(-5)(-4)$$

and therefore $D = 1/2$. This leaves only C to be determined; simply equating the coefficients of the s^3 term gives

$$0 \equiv A + B + C$$

and therefore $C = -A - B = -7/40$.

Method 2

Let

$$\frac{f(s)}{g(s)} = \frac{f(s)}{(s + s_1)(s + s_2)(s + s_3)^k} \qquad (3.55)$$

and the corresponding partial-fraction expansion is

$$\frac{f(s)}{g(s)} = \frac{A}{s + s_1} + \frac{B}{s + s_2} + \frac{C_1}{s + s_3} + \ldots + \frac{C_k}{(s + s_3)^k} \qquad (3.56)$$

where the constants $A, B, C_1, \ldots, C_k$ are called the *residues*; thus

$$\frac{f(s)}{(s + s_1)(s + s_2)(s + s_3)^k} = \frac{A}{s + s_1} + \frac{B}{s + s_2} + \frac{C_1}{s + s_3} +$$

$$\ldots + \frac{C_k}{(s + s_3)^k} \qquad (3.57)$$

Now if both sides are multiplied by $(s + s_1)$ and the resulting identity is evaluated at $s = -s_1$, then

$$A = \left. \frac{f(s)}{(s + s_2)(s + s_3)^k} \right|_{s=s_1} \qquad (3.58)$$

since all the other terms contain a factor $(s + s_1)$ and are therefore zero. Similarly

$$B = \left. \frac{f(s)}{(s + s_1)(s + s_3)^k} \right|_{s=-s_2} \qquad (3.59)$$

Using this procedure in the case of the multiple root, the only residue that can be evaluated is C_k: multiplying both sides of equation 3.57 by $(s + s_3)^k$ and letting $s = -s_3$ gives

$$C_k = \frac{f(s)}{(s + s_1)(s + s_2)} \bigg|_{s=-s_3}$$

To obtain C_1-C_{k-1} multiply both sides of equation 3.57 by $(s + s_3)^k$ as previously, to give

$$\frac{f(s)}{(s + s_1)(s + s_2)} = \frac{A(s + s_3)^k}{s + s_1} + \frac{B(s + s_3)^k}{s + s_2}$$
$$+ C_1(s + s_3)^{k-1} + \ldots + C_{k-1}(s + s_3) + C_k \qquad (3.60)$$

and then differentiate both sides with respect to s and evaluate the resulting identify at $s = -s_3$ to give

$$C_{k-1} = \left[\frac{d}{ds} \left\{ \frac{f(s)}{(s + s_1)(s + s_2)} \right\} \right]_{s=-s_3} \qquad (3.61)$$

Repeated differentiation of both sides of equation 3.60 will give the other residues $C_{k-2}, \ldots, C_1$, in turn.

Example 3.4
Use the residue method to express the following ratio in partial fractions

$$\frac{1}{(s + 1)(s + 5)^3}$$

Let

$$\frac{1}{(s + 1)(s + 5)^3} = \frac{A}{s + 1} + \frac{B_1}{s + 5} + \frac{B_2}{(s + 5)^2} + \frac{B_3}{(s + 5)^3}$$

where

$$A = \frac{1}{(s + 5)^3} \bigg|_{s=-1} = \frac{1}{64}$$

and

$$B_3 = \frac{1}{s + 1} \bigg|_{s=-5} = -\frac{1}{4}$$

To obtain B_2, multiply both sides of the identity by $(s + 5)^3$

$$\frac{1}{s + 1} = \frac{A(s + 5)^3}{s + 1} + B_1(s + 5)^2 + B_2(s + 5) + B_3$$

and then differentiate with respect to s to give

$$-\frac{1}{(s + 1)^2} = A \frac{d}{ds}\left[\frac{(s + 5)^3}{s + 1}\right] + 2B_1(s + 5) + B_2$$

and put $s = -5$ to give

$$-\frac{1}{16} = B_2$$

Differentiate again with respect to s

$$\frac{2}{(s+1)^3} = A\frac{d^2}{ds^2}\left[\frac{(s+5)^3}{s+1}\right] + 2B_1$$

(and note the multiplying factor preceding B_1) and again put $s = -5$ to give

$$-\frac{2}{64} = 2B_1$$

Therefore

$$B_1 = -\frac{1}{64}$$

3.7 SOLUTION OF LINEAR DIFFERENTIAL EQUATIONS USING THE LAPLACE TRANSFORMATION

Consider now how the Laplace transformation can be of assistance in the solution of the nth-order linear differential equation 3.1 assuming *zero* initial conditions. Thus using the results found in equations 3.27–3.29 the Laplace transform of equation 3.1 becomes

$$a_0 s^n X(s) + a_1 s^{n-1} X(s) + \ldots + a_{n-1} s X(s) + a_n X(s)$$
$$= b_0 s^m U(s) + \ldots + b_m U(s) \qquad (3.62)$$

where $X(s)$ and $U(s)$ are the Laplace transforms of $x(t)$ and $u(t)$, respectively. Equation 3.62 is an algebraic equation and may be treated as such to give

$$X(s) = \left[\frac{b_0 s^m + b_1 s^{m-1} + \ldots + b_{m-1} s + b_m}{a_0 s^n + a_1 s^{n-1} + \ldots + a_{n-1} s + a_n}\right] U(s) \quad (n > m) \qquad (3.63)$$

Equation 3.63 represents the transformed relationship between the input $U(s)$ and the output $X(s)$ and the term in the square brackets is called the *transfer function* between the input and output *provided the initial conditions are zero*. The transfer function is of central importance in all control work; the mathematical model of a system is usually expressed in this form (rather than in differential-equation form) provided the system is *linear* and all initial conditions are *zero*. Also, since the transform of the impulse function is unity (see equation 3.38), the transfer function is the transform of the time response that would result if the system were excited by the impulse function.

In exactly the same way as the roots of the auxiliary equation (equation 3.15) determine the system stability, so do the roots of the denominator of the transfer function in equation 3.63. In fact the auxiliary equation and this denominator are identical and have identical roots; this is not too surprising since both forms represent the same system behaviour, albeit in a slightly different way. Thus, for stability the roots of the denominator of the transfer function must have negative real parts.

In order to obtain the time solution $x(t)$ of equation 3.1 the inverse Laplace transform of equation 3.63 must be obtained by

(1) obtaining the transform $U(s)$
(2) factorising the denominator
(3) expressing the resulting ratio of polynomials in partial-fraction form
(4) using the table of Laplace-transform pairs in an inverse manner to obtain the time functions associated with each of the partial fractions and
(5) obtaining the time solution $x(t)$ as the summation of these individual functions.

This procedure is illustrated by the following example.

Example 3.5
Use the Laplace transformation to obtain the solution of the second-order differential equation

$$\frac{d^2 x(t)}{dt^2} + 3 \frac{dx(t)}{dt} + 2x(t) = 2u(t)$$

(1) with zero initial conditions and $u(t)$ the impulse function, and (2) with $x(0) = 2$, $dx(0)/dt = 0$, and $u(t)$ the unit step function.
(1) Obtain the transform using equations 3.27 and 3.28, inserting the appropriate initial conditions (note that this is in direct contrast with the classical method), and equation 3.38

$$s^2 X(s) + 3s X(s) + 2 X(s) = 2$$

Therefore

$$X(s) = \frac{2}{s^2 + 3s + 2} = \frac{2}{(s+1)(s+2)}$$

and expressing in partial-fraction form gives

$$X(s) = \frac{A}{s+1} + \frac{B}{s+2}$$

where $A = 2$ and $B = -2$ using the residue method. Thus

$$X(s) = \frac{2}{s+1} - \frac{2}{s+2}$$

and using the table of transform pairs

$$x(t) = 2e^{-t} - 2e^{-2t}$$

which is the solution of the differential equation.
(2) Again obtain the transform using equations 3.27 and 3.28 and also equation 3.39

$$s^2 X(s) - 2s + 3s X(s) - 3.2 + 2X(s) = 2\frac{1}{s}$$

Therefore
$$(s^2 + 3s + 2)X(s) = 2s + 6 + \frac{2}{s} = \frac{2s^2 + 6s + 2}{s}$$

and
$$X(s) = \frac{2s^2 + 6s + 2}{s(s+1)(s+2)}$$

Expressing in partial-fraction form gives
$$X(s) = \frac{1}{s} + \frac{2}{s+1} - \frac{1}{s+2}$$

and therefore
$$x(t) = 1 + 2e^{-t} - e^{-2t}$$

which is of similar form to the result previously obtained. Note that the responses are stable.

3.8 SECOND-ORDER DIFFERENTIAL EQUATIONS

In the previous chapter, mathematical models of a mechanical system containing spring, mass and dashpot, and an electrical system containing resistance, induction and capacitance were obtained in the form of second-order linear differential equations. Later on it will become evident that, in many cases of analysis, higher-order systems may be approximately represented by second-order equations and hence it is important to have a full understanding of the behaviour of this equation.

Consider the general second-order differential equation
$$\frac{d^2 x(t)}{dt^2} + 2\xi\omega_n \frac{dx(t)}{dt} + \omega_n^2 x(t) = \omega_n^2 u(t) \tag{3.64}$$

where $x(t)$ is the output, $u(t)$ is the input, ξ is the *damping factor*, and ω_n is the *undamped natural frequency*. The meaning of these terms will become evident later on.

Thus in the case of the mechanical system
$$2\xi\omega_n = \frac{R}{M}; \quad \omega_n^2 = \frac{K}{M}$$

and for the electrical system
$$2\xi\omega_n = \frac{R}{L}; \quad \omega_n^2 = \frac{1}{LC}$$

Assuming zero initial conditions, the transform of equation 3.64 is
$$s^2 X(s) + 2\xi\omega_n s X(s) + \omega_n^2 X(s) = \omega_n^2 U(s) \tag{3.65}$$

that is

$$X(s) = \left[\frac{\omega_n^2}{s^2 + 2\xi\omega_n s + \omega_n^2}\right] U(s) \qquad (3.66)$$

where the term in square brackets is the transfer function. Taking the input to be the unit step function

$$X(s) = \frac{\omega_n^2}{s(s^2 + 2\xi\omega_n s + \omega_n^2)} \qquad (3.67)$$

that is

$$X(s) = \frac{\omega_n^2}{s[s + \xi\omega_n + \sqrt{(\xi^2 - 1)}\omega_n][s + \xi\omega_n - \sqrt{(\xi^2 - 1)}\omega_n]} \qquad (3.68)$$

As can be seen, depending upon whether $\xi > 1$, $\xi = 1$, or $\xi < 1$, the roots of the denominator will be real and unequal, real and equal, or complex in the form of a conjugate pair. Each of these possibilities will be discussed in turn.

Real and Unequal Roots ($\xi > 1$)

In this case

$$X(s) = \frac{1}{s} + \frac{A}{s + s_1} + \frac{B}{s + s_2}$$

where

$$s_1 = \xi\omega_n + \sqrt{(\xi^2 - 1)}\omega_n$$
$$s_2 = \xi\omega_n - \sqrt{(\xi^2 - 1)}\omega_n$$
$$A = \frac{1}{2\xi\sqrt{(\xi^2 - 1)} + 2(\xi^2 - 1)}$$
$$B = \frac{1}{-2\xi\sqrt{(\xi^2 - 1)} + 2(\xi^2 - 1)}$$

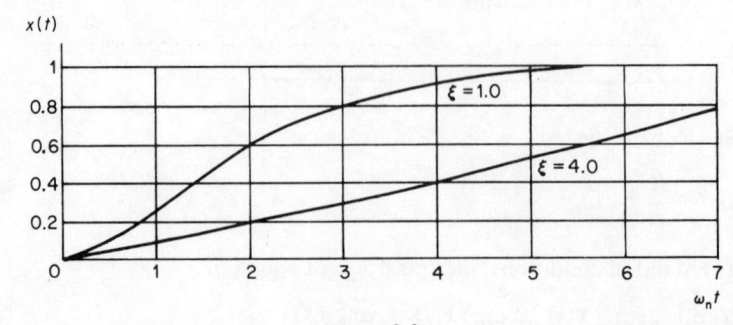

Figure 3.3

The Need for Some Mathematics

Taking inverse Laplace transforms leads to the result

$$x(t) = 1 + Ae^{-s_1 t} + Be^{-s_2 t} \tag{3.69}$$

which is shown in figure 3.3 for two values of the damping factor ξ. It is seen that, as ξ increases, the response becomes more sluggish, which in the case of the mechanical system is equivalent to increasing the viscous-friction effect whilst keeping the other variables constant. When $\xi > 1$ the system is said to be overdamped.

Equal Roots ($\xi = 1$)

In this case

$$X(s) = \frac{\omega_n^2}{s(s + \omega_n)^2} = \frac{1}{s} - \frac{1}{s + \omega_n} - \frac{\omega_n}{(s + \omega_n)^2}$$

and (from the table of transform pairs) the inverse Laplace transform is

$$x(t) = 1 - e^{-\omega_n t} - \omega_n t e^{-\omega_n t}$$
$$= 1 - e^{-\omega_n t}(1 + \omega_n t) \tag{3.70}$$

which is shown in figure 3.3. When $\xi = 1$, the system is said to be critically damped.

Complex-conjugate Roots ($\xi < 1$)

From equation 3.67

$$X(s) = \frac{\omega_n^2}{s(s^2 + 2\xi\omega_n s + \omega_n^2)}$$

The roots of the denominator term in brackets are $s_1, s_2 = -\xi\omega_n \pm j\sqrt{(1 - \xi^2)}\omega_n$, which are shown plotted in figure 3.4. Note that $\cos \phi = \xi$, the damping factor.

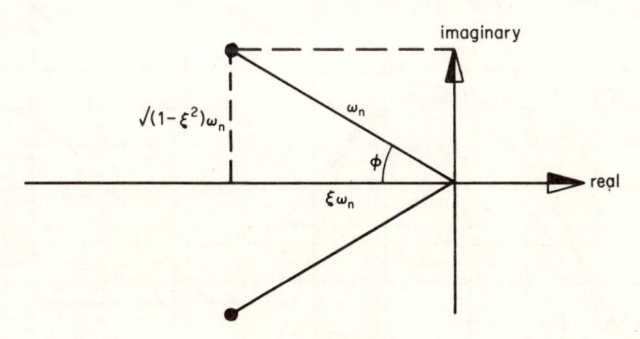

Figure 3.4

Now

$$X(s) = \frac{\omega_n^2}{s[(s + \xi\omega_n)^2 + (1 - \xi^2)\omega_n^2]}$$

Introduction to Control Theory

$$= \frac{1}{s} - \frac{s + 2\xi\omega_n}{(s + \xi\omega_n)^2 + (1 - \xi^2)\omega_n^2}$$

$$= \frac{1}{s} - \frac{s + \xi\omega_n}{(s + \xi\omega_n)^2 + (1 - \xi^2)\omega_n^2} - \frac{\xi\omega_n}{(s + \xi\omega_n)^2 + (1 - \xi^2)\omega_n^2}$$

and using the table of transform pairs

$$x(t) = 1 - e^{-\xi\omega_n t} \cos\sqrt{(1 - \xi^2)}\omega_n t - e^{-\xi\omega_n t} \frac{\xi\omega_n}{\sqrt{(1-\xi^2)}\omega_n} \sin\sqrt{(1 - \xi^2)}\omega_n t$$

$$= 1 - e^{-\xi\omega_n t} \cos\sqrt{(1 - \xi^2)}\omega_n t - \frac{\xi}{\sqrt{(1-\xi^2)}} \sin\sqrt{(1 - \xi^2)}\omega_n t \tag{3.71}$$

$$= 1 - \frac{1}{\sqrt{(1 - \xi^2)}} e^{-\xi\omega_n t} \sin(\omega_n\sqrt{(1 - \xi^2)}t + \phi) \tag{3.72}$$

where

$$\phi = \tan^{-1} \frac{\sqrt{(1 - \xi^2)}}{\xi}$$

This response is shown in figure 3.5 for values of the damping factor ξ between 0 and 1; as is evident, it is this damping factor that dominates the response whilst ω_n merely affects the time scale. When $\xi = 0$ (no damping), the response is a sinusoid of natural frequency ω_n, the undamped natural frequency. As ξ increases the curves become less oscillatory with natural frequency $\omega_r = \omega_n\sqrt{(1 - \xi^2)}$, known

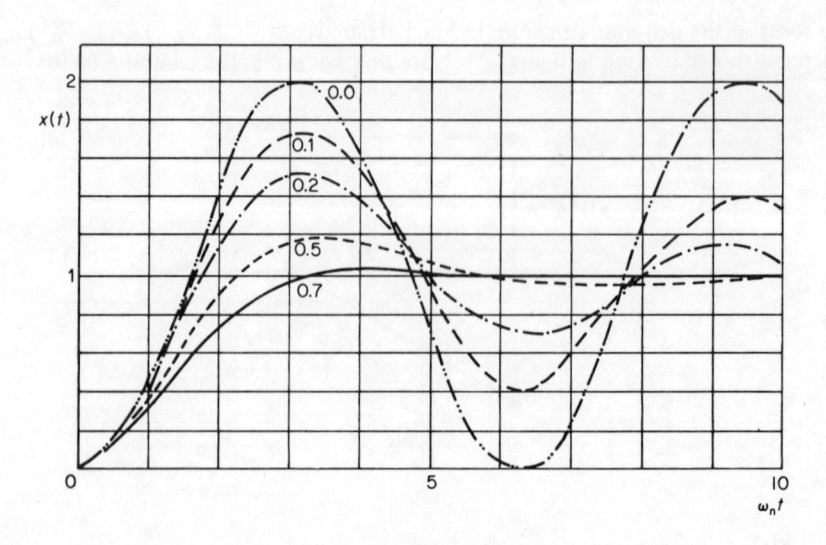

Figure 3.5

The Need for Some Mathematics

as the *damped frequency*, and the percentage overshoot decreases. This latter term is defined by

$$\text{Percentage overshoot} = 100\left(\frac{\text{maximum value of } x(t) - \text{steady-state value}}{\text{steady-state value}}\right) \quad (3.73)$$

and is calculated as follows. At a peak value, $dx(t)/dt = 0$, and therefore from equation 3.72

$$\frac{dx(t)}{dt} = \frac{\xi\omega_n}{\sqrt{(1-\xi^2)}} e^{-\xi\omega_n t} \sin(\omega_r t + \phi) - \omega_n e^{-\xi\omega_n t} \cos(\omega_r t + \phi) = 0 \quad (3.74)$$

when

$$\tan(\omega_r t + \phi) = \frac{\sqrt{(1-\xi^2)}}{\xi}$$

that is

$$\omega_r t + \phi = \tan^{-1}\frac{\sqrt{(1-\xi^2)}}{\xi}$$

and therefore

$$\omega_r t = n\pi$$

or

$$t = \frac{n\pi}{\omega_r} = \frac{n\pi}{\omega_n\sqrt{(1-\xi^2)}} \quad , n = 0, 1, 2, 3, \ldots \quad (3.75)$$

The maximum value of $x(t)$ occurs when $n = 1$ at time

$$t_{max} = \frac{\pi}{\omega_n\sqrt{(1-\xi^2)}} \quad (3.76)$$

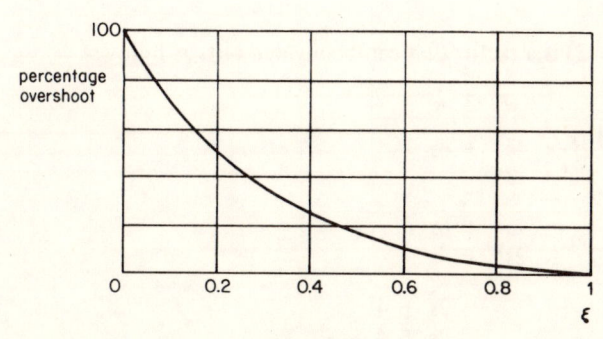

Figure 3.6

(equal to half the period of oscillation associated with the damped frequency) and, from equation 3.72, its value is

$$x_{max} = 1 + e^{-[\xi\pi/\sqrt{(1-\xi^2)}]} \tag{3.77}$$

Therefore the percentage overshoot is

$$100e^{-[\xi\pi/\sqrt{(1-\xi^2)}]} \tag{3.78}$$

and is shown as a function of ξ in figure 3.6.

Finally, when $\xi < 1$, the system is said to be underdamped.

Example 3.6
Find the solution, using Laplace transformation methods, of the differential equation

$$\frac{d^3x(t)}{dt^3} + 4\frac{d^2x(t)}{dt^2} + 6\frac{dx(t)}{dt} + 4x(t) = 4u(t)$$

where $u(t)$ is the unit step function and all initial conditions are zero.

Taking Laplace transforms gives

$$s^3 X(s) + 4s^2 X(s) + 6sX(s) + 4X(s) = \frac{4}{s}$$

Therefore

$$X(s) = \frac{4}{s(s^3 + 4s^2 + 6s + 4)}$$

It is now necessary to factorise the denominator and, since all the coefficients are positive, the real root must be negative. Thus, try a few values: for $s_1 = 0$

$$f(s_1) = 4$$

for $s_1 = -1$

$$f(s_1) = 1$$

for $s_1 = -2$

$$f(s_1) = 0$$

Therefore $(s + 2)$ is a factor that can be divided out, as follows

$$\begin{array}{r} s^2 + 2s + 2 \\ s + 2 \overline{\smash{\big)}\ s^3 + 4s^2 + 6s + 4} \\ \underline{s^3 + 2s^2} \\ 2s^2 + 6s \\ \underline{2s^2 + 4s} \\ 2s + 4 \\ \underline{2s + 4} \end{array}$$

to give $s^3 + 4s^2 + 6s + 4 = (s + 2)(s^2 + 2s + 2)$ and it can easily be verified that the quadratic factor has complex roots $-1 \pm j$. Therefore

$$X(s) = \frac{4}{s(s+2)(s^2+2s+2)} = \frac{A}{s} + \frac{B}{s+2} + \frac{Cs+D}{s^2+2s+2}$$

where $A = 1$, $B = -1$ using the residue method, and therefore

$$4 \equiv (s+2)(s^2+2s+2) - s(s^2+2s+2) + (Cs+D)s(s+2)$$

Equating coefficients of the s^3 term gives

$$0 \equiv 1 - 1 + C$$

and therefore $C = 0$. Equating coefficients of the s term gives

$$0 \equiv 6 - 2 + 2D$$

and therefore $D = -2$. Thus

$$X(s) = \frac{1}{s} - \frac{1}{s+2} - \frac{2}{s^2+2s+2}$$

$$= \frac{1}{s} - \frac{1}{s+2} - \frac{2}{(s+1)^2+1}$$

and taking inverse Laplace transforms

$$x(t) = 1 - e^{-2t} - 2e^{-t}\sin t$$

REFERENCES

Gardner, M. F., and Barnes, J. L., *Transients in Linear Systems* (Wiley, New York, 1942).

PROBLEMS

3.1 Obtain the Laplace transforms of the following functions from first principles

   (i) $\cos \omega t$
   (ii) $at - 1 - e^{-at}$
   (iii) $\cosh at$.

3.2 Express the following ratios in partial-fraction form

   (i) $\dfrac{1}{s^3 + 3s^2 + 2.75s + 0.75}$

   (ii) $\dfrac{s+2}{s^3 + 3s^2 + 2.75s + 0.75}$

   (iii) $\dfrac{1}{s^5 + 4s^4 + 9s^3 + 12s^2 + 10s + 4}$

3.3 Obtain the solution of the differential equation

$$\frac{d^2x(t)}{dt^2} + \frac{2dx(t)}{dt} + 5x(t) = 5u(t)$$

(i) with $u(t)$ the unit step function and all initial conditions zero;
(ii) with $u(t)$ the unit step function and $x(0) = 1$, $dx(0)/dt = 0$;
(iii) with $u(t) = e^{-2t}$ and all initial conditions zero.

3.4 Obtain the mathematical model of the spring–dashpot–mass system shown in figure 2.5 when $M = 1$, $R = 2$ and $K = 2$ (consistent units being assumed), and determine the behaviour of the model if $u(t)$ is a unit step function and all initial conditions are zero.

3.5 Obtain the mathematical model of the electrical system in figure 3.7. Determine the behaviour of the output voltage $V_o(t)$, given that the applied voltage $V_i(t)$ is a step function of 100 V and assuming that all initial conditions are zero.

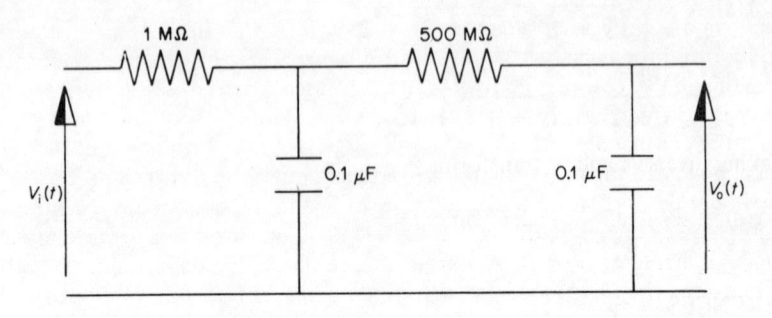

Figure 3.7

4 Control-system Representation

4.1 OPEN-LOOP AND CLOSED-LOOP CONTROL

The control of a process or system may be achieved either in open loop or in closed loop and the concepts of both were discussed in the introductory chapter. Before going on to consider control-system representation, it is necessary to quantify the differences between open-loop and closed-loop control; to this end, consider the control of the boiler steam pressure in a boiler-turbine unit delivering electrical power to the grid system, shown diagrammatically in figure 4.1.

The boiler delivers steam at constant pressure and temperature at the turbine governor valve (TGV) by burning fuel such as coal, oil or gas in a furnace, thus producing radiant heat, which evaporates the circulating water forming a steam-water mixture. The steam separates out in the boiler drum, and flows through the superheaters to the TGV. Arriving at the turbine, it is expanded, thereby causing the turbine to rotate at a certain speed.

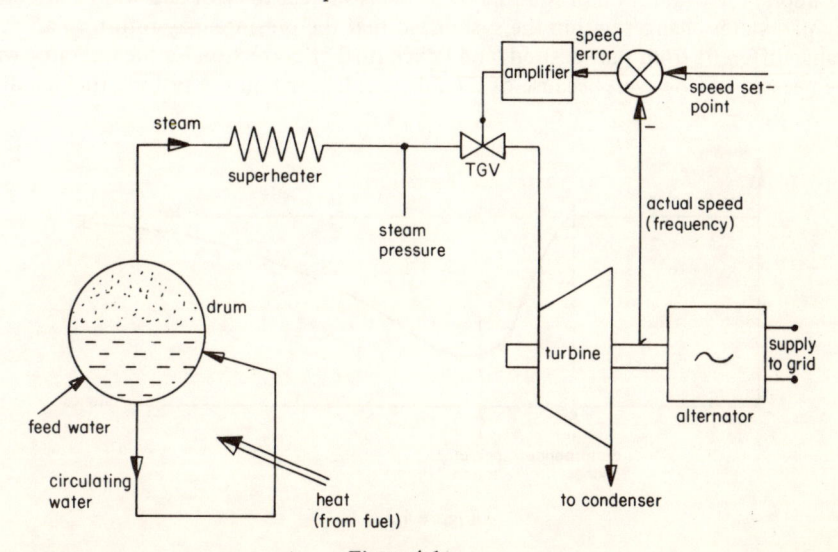

Figure 4.1

It is assumed that the turbine speed (or frequency) is controlled by using a Watt flyball governor to open and close the TGV, thus admitting more or less steam to the turbine. This TGV action causes the steam pressure to vary, but for various design reasons a constant steam pressure is required. This is achieved by the manipulation of the fuel feed rate, which leads to changes in the amount of heat generated in the furnace and hence the boiler steam pressure. Thus, figure 4.1 can be simplified by assuming that the signal to the TGV is a disturbance input as shown, again diagrammatically, in figure 4.2.

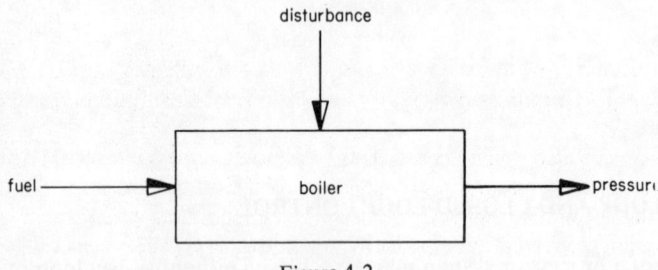

Figure 4.2

Consider now the control of steam pressure at its design value in an open-loop manner: in this case, following a disturbance on the boiler that causes the pressure to drop, the operator increases the fuel feed rate by a known amount so as to restore the pressure to its design value or thereabouts, as shown in figure 4.3. Here it can be seen that, because of the *thermal inertia* of the boiler system, the pressure cannot be made to change instantaneously, and following operator intervention it takes an appreciable time for the pressure to be restored.

This form of control is extremely simple, and may well operate satisfactorily in many applications, but it will obviously suffer from many disadvantages. For example, the calibration of the fuel feed rate is subject to error and will be affected by parameter changes within the system, so that the pressure may return to a value different from that desired, and hence further correction by the operator will be necessary. However, because of the inherent lags within the system, the operator

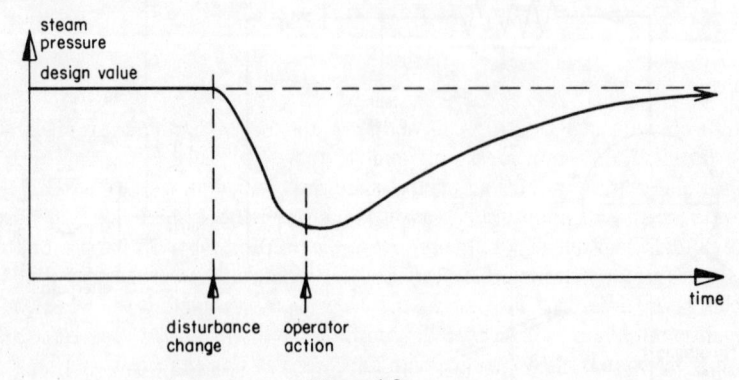

Figure 4.3

Control-system Representation

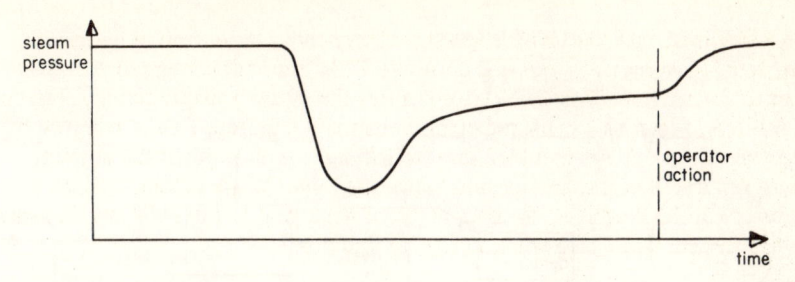

Figure 4.4

can only apply such correction once the effects of the initial change are known, as shown in figure 4.4, and thus the pressure will remain offset for a considerable period of time.

The procedure is also complicated if, for example, a further disturbance change takes place before the pressure has settled out, as shown in figure 4.5. Thus, if the pressure has to be maintained at a constant value in the presence of frequent disturbance changes, system-parameter changes, etc., it is evident that this can only be done by using some automatic system—in other words, some form of closed-loop control system. That is, the steam pressure is continually measured and compared with its design value or set-point and the error between the two is used, in some manner, to control the amount of fuel being burnt in the furnace. This is the feature of all closed-loop control systems: the output or the error in the output is used to determine the input in such a way that the error is reduced to

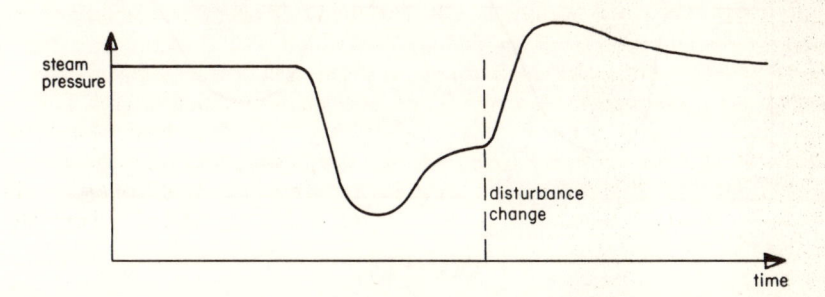

Figure 4.5

zero. This is shown in figure 4.6. A closed-loop control system can, provided it is correctly designed, maintain pressure equal to its desired value following both disturbance and system-parameter changes; such systems are used extensively throughout industry to maintain certain variables at pre-designed values.

The main disadvantage (apart from cost and complexity) is that if the control system is *not* correctly designed it can exhibit instability due to the fact that the system being controlled has inertia, that is, there is a time lag between cause and effect, as shown in figure 4.3. Thus, for example, following a drop in steam pressure, the fuel may be increased by too great an amount, causing the pressure to overshoot its desired value and this in turn will cause the fuel to decrease, and so

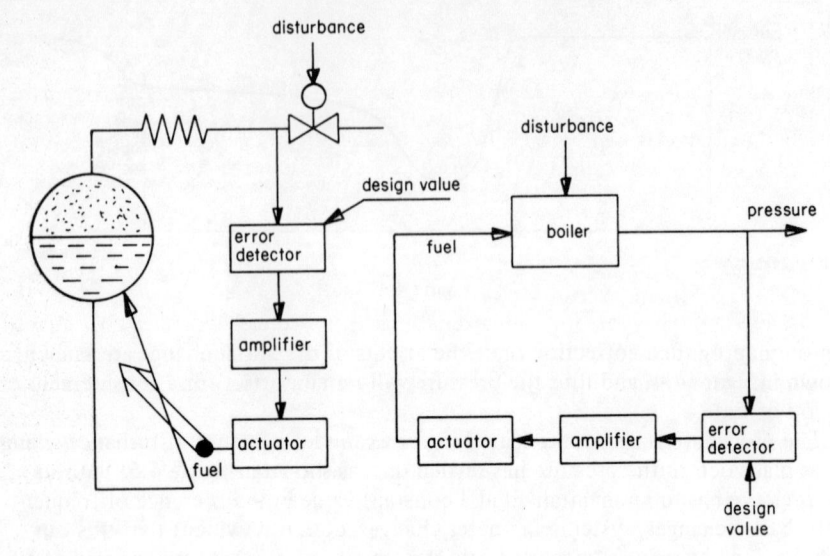

Figure 4.6

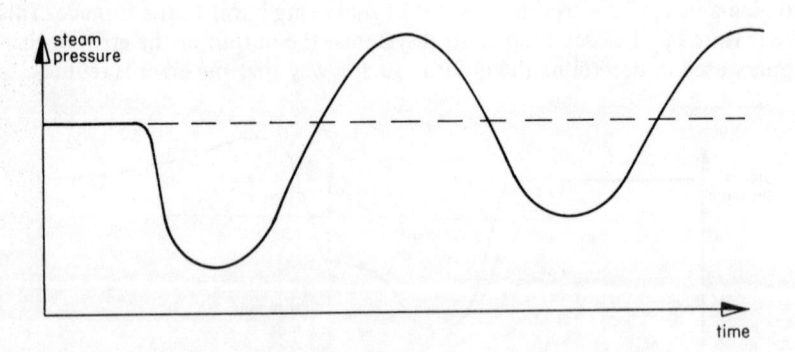

Figure 4.7

on; this situation is shown in figure 4.7. Thus, it is all too easy to see how an instability can be set up as a result of bad control-system design.

In general, a closed-loop control system will take the form shown diagrammatically in figure 4.8. If the reference value (or design value or set-point) is constant at some pre-determined value, and the aim of the control system is to maintain the output equal to this reference value following changes in system parameters or disturbances, the control system is termed a *regulator*. If, on the other hand, the output has to follow changes in the reference value over a wide range, the system is termed a *servomechanism*.

The analysis and design of closed-loop control systems such that their performance and stability requirements are achieved is the main subject matter of this book, and subsequent chapters consider the various methods that are

Control-system Representation

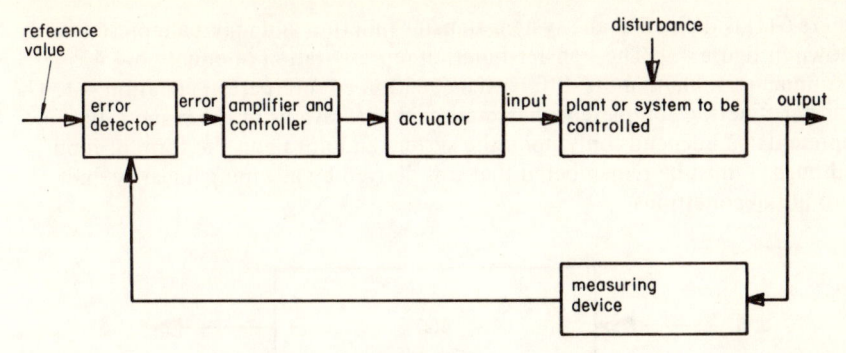

Figure 4.8

available to engineers to enable them to do this. First, however, it is necessary to consider the representation of such systems in the form of transfer functions, block diagrams and signal-flow graphs.

4.2 TRANSFER FUNCTIONS

In chapter 2, mathematical models of various systems were developed by considering the physics and chemistry of these systems and writing down the relevant equations of motion. This led to models consisting of non-linear differential equations, which, it was stated, could be linearised. The argument advanced for linearisation was that most control-system design techniques are based on linear theory, and that systems under closed-loop control should experience only small perturbations about some reference value, and hence can be considered to behave in a linear fashion about that value.

This procedure results, in general, in a set of linear differential equations connecting the inputs and outputs of the system, but in this book (except for the final chapter) only systems with one input and one output and perhaps a disturbance input are considered.

Thus the mathematical model relating the input and output is

$$a_0 \frac{d^n x}{dt^n} + \ldots + a_n x = b_0 \frac{d^m u}{dt^m} + \ldots + b_m u \tag{4.1}$$

where $n \geq m + 1$, the condition for the system to be realisable. Taking Laplace transforms gives, assuming all initial conditions to be zero, the result

$$a_0 s^n x(s) + \ldots + a_n x(s) = b_0 s^m u(s) + \ldots + b_m u(s) \tag{4.2}$$

that is

$$x(s) = \frac{b_0 s^m + \ldots + b_m}{a_0 s^n + \ldots + a_n} u(s) \tag{4.3}$$

$$= \frac{f(s)}{g(s)} u(s) = G(s) u(s) \tag{4.4}$$

where $G(s)$ is defined as the system transfer function and may be represented as shown in figure 4.9. The transfer-function representation of equation 4.4 is far less unmanageable than its differential-equation counterpart in equation 4.1. The transfer function completely characterises the behaviour of the system that it represents and depends only upon the system and not upon the form of input, although it must be remembered that it is derived by assuming linearity and zero initial conditions.

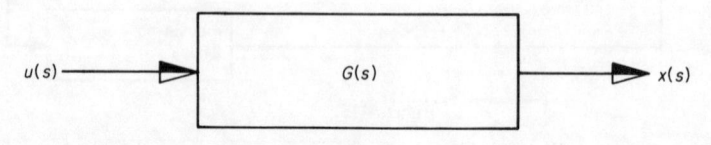

Figure 4.9

It must be mentioned that not all transfer functions are rational algebraic expressions as given by equation 4.3, since a transfer function of a system that is simply a pure time delay is e^{-sT}, where T is the value of that delay. For example, consider a fluid flowing through a pipe where $q_i(t)$ is the inlet flow and $q_o(t)$ is the outlet flow and the time required for the fluid to flow along the pipe is T. Then the relationship between the flows is

$$q_o(t) = q_i(t - T) \tag{4.5}$$

as shown in figure 4.10. Taking Laplace transforms gives

$$q_o(s) = e^{-sT} q_i(s) \tag{4.6}$$

using equation 3.32 and therefore the transfer function is

$$\frac{q_o(s)}{q_i(s)} = e^{-sT} \tag{4.7}$$

which is the Laplace transform of the pure time delay.

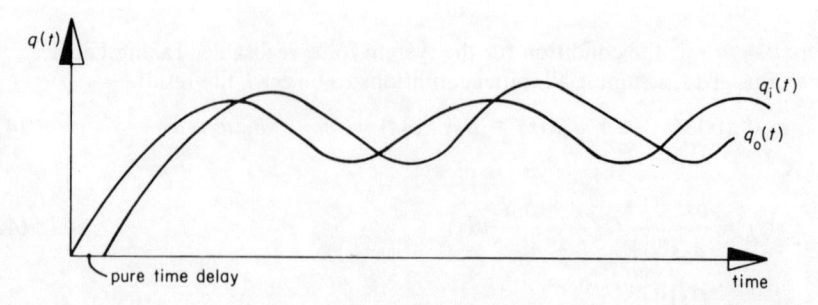

Figure 4.10

Example 4.1

The transfer function of a system between input $u(t)$ and output $x(t)$, is $G_1(s)$ and that between disturbance $d(t)$ and output $x(t)$ is $G_2(s)$. Then, since the system is linear, the principle of superposition holds and the complete relationship is

$$x(s) = G_1(s)u(s) + G_2(s)d(s)$$

which is shown in figure 4.11. Identify the input, output and disturbance variables with respect to the steam boiler discussed earlier.

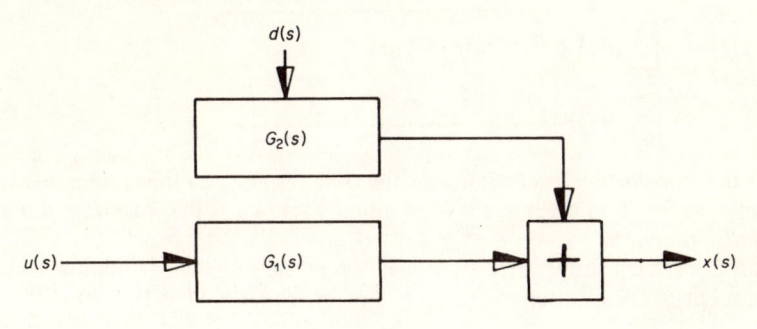

Figure 4.11

Stability and Steady-state Gain

In the previous chapter it was stated that the stability of a linear system is determined by the characteristic equation of equation 3.1, which is identical to setting the denominator of the system transfer function equal to zero

$$g(s) = 0 \tag{4.8}$$

For stability the roots of this equation must have negative real parts and are, in general, complex, occurring as conjugate pairs. These roots are called the *poles* of $G(s)$; the roots of $f(s)$, the numerator of $G(s)$, are called the *zeros* of $G(s)$; much more will be said about these parameters in chapter 6.

Now, by using the final-value theorem (equation 3.37) and assuming that the input is the unit step function, the steady-state gain K_{ss} of the system (provided it exists, that is, providing the system is stable) is defined as

$$K_{ss} = \frac{b_m}{a_n} \tag{4.9}$$

that is

$$x(\infty) = x_{ss} = \frac{b_m}{a_n} \tag{4.10}$$

Convolution Integral

Consider the system shown in figure 4.9, where

$$x(s) = G(s)u(s)$$

Introduction to Control Theory

and therefore

$$x(t) = \mathcal{L}^{-1}[G(s)u(s)] \qquad (4.11)$$
$$= \mathcal{L}^{-1}\left[G(s)\int_0^\infty u(\tau)e^{-s\tau}d\tau\right] \qquad (4.12)$$

using the definition of Laplace transformation. It can be shown that the operations of integration and inverse transformation can be interchanged and hence equation 4.12 can be written as

$$x(t) = \int_0^\infty u(\tau)\mathcal{L}^{-1}[G(s)e^{-s\tau}]d\tau \qquad (4.13)$$
$$= \int_0^\infty u(\tau)w(t-\tau)d\tau \qquad (4.14)$$

This is the *convolution integral*; $w(t)$ is the time response of the system due to a unit-impulse-function input at $t = 0$, commonly known as the *weighting function* or *impulse function*.

Thus it has been shown that multiplication in the s domain—equation 4.4,— is equivalent to convolution in the time domain—equation 4.14.

Example 4.2

To illustrate in some detail how to obtain the transfer function of a system from first principles, consider the inverted pendulum shown in figure 4.12, where $u(t)$

Figure 4.13

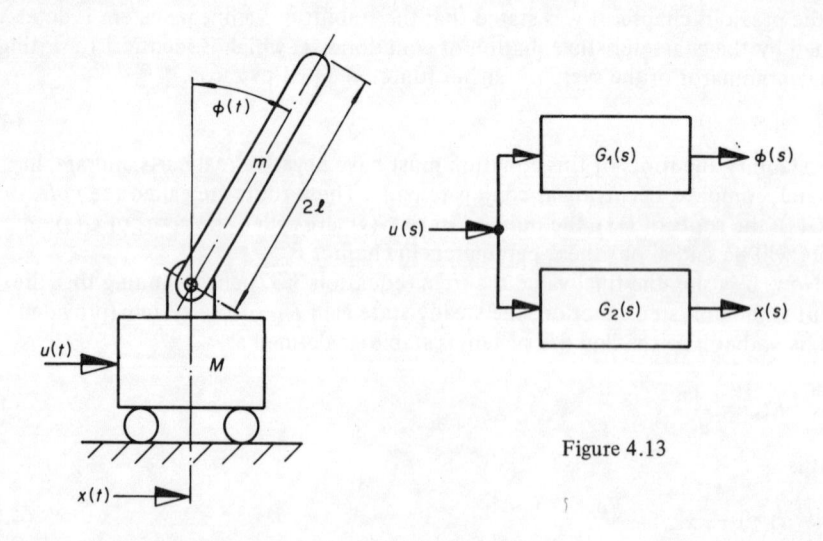

Figure 4.12

is the input force, $\phi(t)$ is the angular position of the pendulum and $x(t)$ is the position of the carriage. Then applying Newton's second law of motion to the pendulum and carriage in turn (assuming that the hinge and wheels are frictionless), the differential equations representing the system may be obtained as

$$I \frac{d^2\phi}{dt^2} = mgl\sin\phi + ml\sin\phi \frac{d^2}{dt^2}(l\cos\phi)$$
$$- ml\cos\phi \frac{d^2}{dt^2}(x + l\sin\phi)$$

and

$$M \frac{d^2x}{dt^2} = u - m \frac{d^2}{dt^2}(x + l\sin\phi)$$

It is clear that these equations are non-linear in ϕ and so, if this variable is assumed to remain small at all times, that is $\sin\phi = \phi$ and $\cos\phi = 1$, and if the second-order terms are neglected, the equations can be effectively linearised and may be written as

$$(ml^2 + I)\frac{d^2\phi}{dt^2} = mgl\phi - ml\frac{d^2x}{dt^2}$$

and

$$(M + m)\frac{d^2x}{dt^2} = u - ml\frac{d^2\phi}{dt^2}$$

Applying the Laplace transformation (assuming zero initial conditions) gives

$$(ml^2 + I)s^2\phi(s) = mgl\phi(s) - mls^2 x(s)$$
$$(M + m)s^2 x(s) = u(s) - mls^2 \phi(s)$$

and therefore after some manipulation

$$\phi(s) = \frac{-ml/(M + m)}{\left(I + \dfrac{Mml^2}{M + m}\right)s^2 - mgl} u(s) = G_1(s)u(s)$$

$$x(s) = \frac{Js^2 - mgl}{(M + m)s^2 \left[\left(I + \dfrac{Mml^2}{M + m}\right)s^2 - mgl\right]} u(s) = G_2(s)u(s)$$

where

$$J = I + \frac{mMl^2 + m^2 l^2}{M + m}$$

The transfer functions are shown in figure 4.13.

4.3 BLOCK-DIAGRAM REPRESENTATION

Figures 4.9 and 4.11 and 4.13 are referred to as *block diagrams*; they are a simple and convenient way of representing the relationships between inputs, disturbances and outputs of linear systems in transfer-function form. Figure 4.9 shows a single block diagram, whilst figures 4.11 and 4.13 represent systems that are a little more complicated. The information path within a block diagram is *one-way only*, from input to output, as indicated by the arrows; the arrows are an important part of the block diagram, and it is good practice to draw them touching the blocks, as shown in the figures.

Complicated systems may be represented by many blocks connected together, each block representing the transfer function of an individual part of the system. Thus, the connection of blocks will represent the *structure* of the complex system and the mathematical relation within each of the blocks—that is, the transfer function connecting the particular input to the output—will represent its mechanism. This interpretation is that proposed in chapter 2 when dealing with mathematical modelling.

Thus, a mathematical model, a transfer function and a block diagram are three identical ways of expressing the input—output information of a linear system.

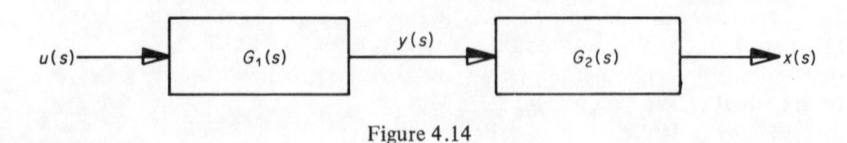

Figure 4.14

Consider the cascaded system shown in figure 4.14; if there is no *interaction* or *loading* between the blocks, that is, if the individual transfer functions remain unaltered when the blocks are linked together, $y(s) = G_1(s)u(s)$ and $x(s) = G_2(s)y(s)$ and hence

$$x(s) = G_2(s)G_1(s)u(s) \tag{4.15}$$

and the over-all transfer function is $G_2(s)G_1(s)$. Although the order of the blocks is not important as far as equation 4.15 is concerned, it is advisable to adhere to this order convention, which indicates that the input passes though $G_1(s)$ first and then $G_2(s)$.

In general, for n blocks in cascade

$$x(s) = G_n(s)G_{n-1}(s) \ldots G_2(s)G_1(s)u(s) \tag{4.16}$$

Block Diagrams of Closed-loop Control Systems

As pointed out earlier, a servomechanism is an example of a closed-loop control system in which the output has to follow changes in the reference value; it may be represented by the block diagram of figure 4.15, which should be compared with figure 4.8. In figure 4.15 $r(s)$ is the transform of the reference value $r(t)$; $x(s)$ is the transformed output; $u(s)$ is the transformed input; $e(s)$ is the error (for correcting the system); $z(s)$ is the output from the measuring device; $K(s)$ is the transfer function of the amplifier and controller; $G(s)$ is the system transfer function; and $H(s)$ is the transfer function of the measuring device.

The plant and measuring device (or transducer) are usually considered to be fixed elements, that is, the transfer functions $G(s)$ and $H(s)$ cannot be changed by the control engineer, whilst the controller is not fixed; it is the aim of control

Control-system Representation

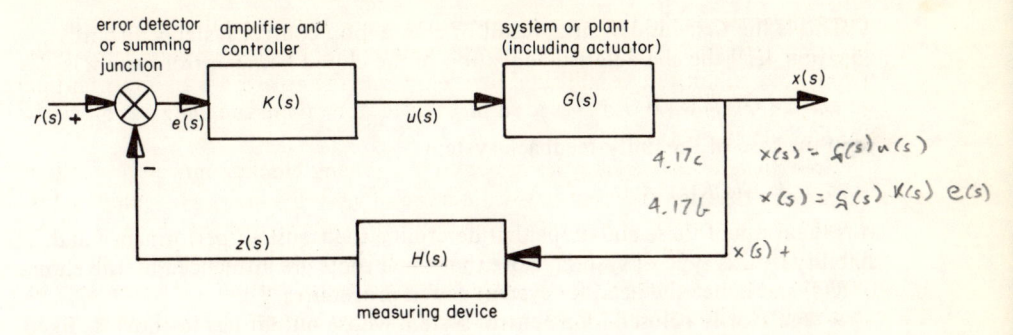

Figure 4.15

engineers to choose the transfer function $K(s)$ such that the closed-loop control system behaves in some prescribed manner. The analysis and design of such systems and, in particular, the choosing of $K(s)$ is the subject of the remaining chapters of this book. It should be noted that the lines of the block diagram represent information flow and not energy flow and although energy is used within the system, as with the steam boiler, it is not shown explicitly and its presence is assumed.

Thus from figure 4.15, the following relationships can be written down: for the summing junction

$$e(s) = r(s) - z(s) \tag{4.17a}$$

(the signs are taken from the input lines; the plus sign is usually omitted); for the controller

$$u(s) = K(s)e(s) \tag{4.17b}$$

for the system or plant

$$x(s) = G(s)u(s) \tag{4.17c}$$

for the measuring device

$$z(s) = H(s)x(s) \tag{4.17d}$$

Combining these equations, the *over-all* input–output relationship can be determined as

$$x(s) = \frac{G(s)K(s)}{1 + G(s)K(s)H(s)} r(s) \tag{4.18}$$

and the over-all transfer function of the closed-loop control system is

$$\frac{x(s)}{r(s)} = G'(s) = \frac{G(s)K(s)}{1 + G(s)K(s)H(s)} \tag{4.19}$$

Assuming *unity* feedback, that is, $H(s) = 1$, then

$$G'(s) = \frac{G(s)K(s)}{1 + G(s)K(s)} \tag{4.20}$$

This is the *first* fundamental result of closed-loop control systems. From equation 4.19 the characteristic equation of the closed-loop control system is

$$1 + G(s)K(s)H(s) = 0 \tag{4.21}$$

or in the case of the unity-feedback system

$$1 + G(s)K(s) = 0 \tag{4.22}$$

It is the roots of these equations that determine the transient performance and stability of this type of system. Note that these roots are influenced by the choice of $K(s)$ and hence the need for system design procedures.

A regulator is a closed-loop control system whose output has to equal its fixed reference value (usually taken to be zero) at all times in the presence of disturbances; the corresponding block diagram is shown in figure 4.16.

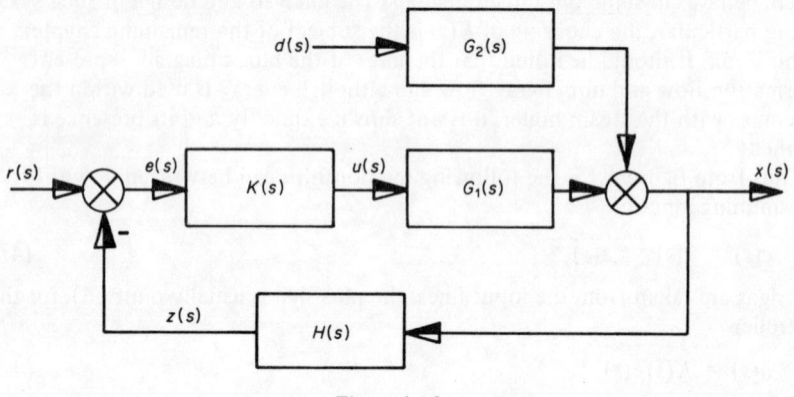

Figure 4.16

The over-all transfer function, with $r(s) \equiv 0$, is

$$x(s) = \frac{G_2(s)}{1 + G_1(s)K(s)H(s)} d(s) \tag{4.23}$$

that is

$$G'(s) = \frac{G_2(s)}{1 + G_1(s)K(s)H(s)} \tag{4.24}$$

and its characteristic equation is unchanged.

Block-diagram Algebra

To assist in the simplification of multi-loop and interlinked loop systems it is helpful to employ a step-by-step procedure sometimes referred to as *block-diagram algebra*. (A more mathematical alternative, the procedure known as *signal-flow graphs*, is presented in the next section.) This block-diagram algebra is

Control-system Representation

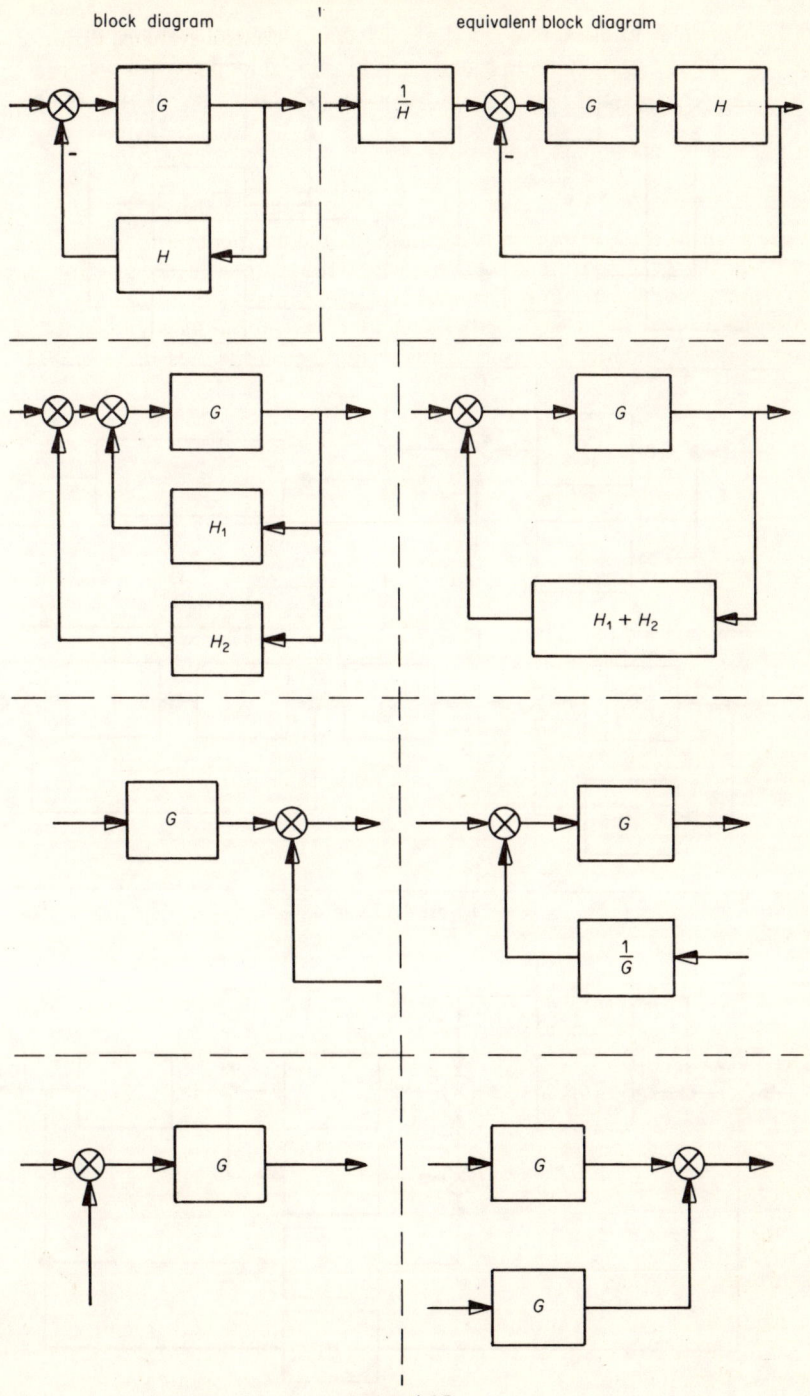

Figure 4.17

Introduction to Control Theory

block diagram | equivalent block diagram

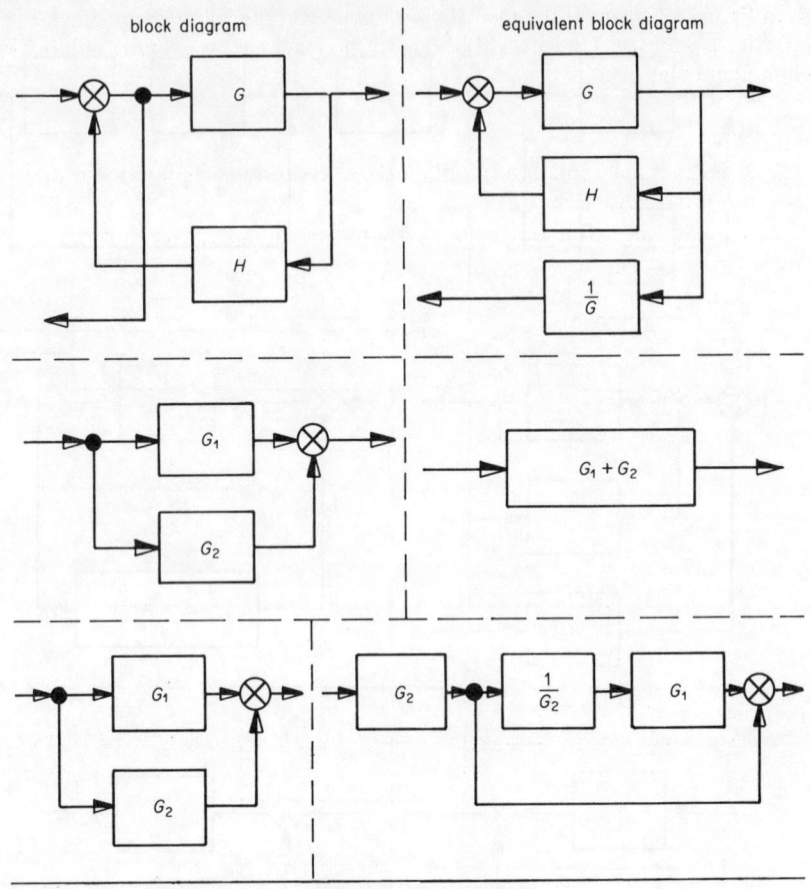

Figure 4.17 contd.

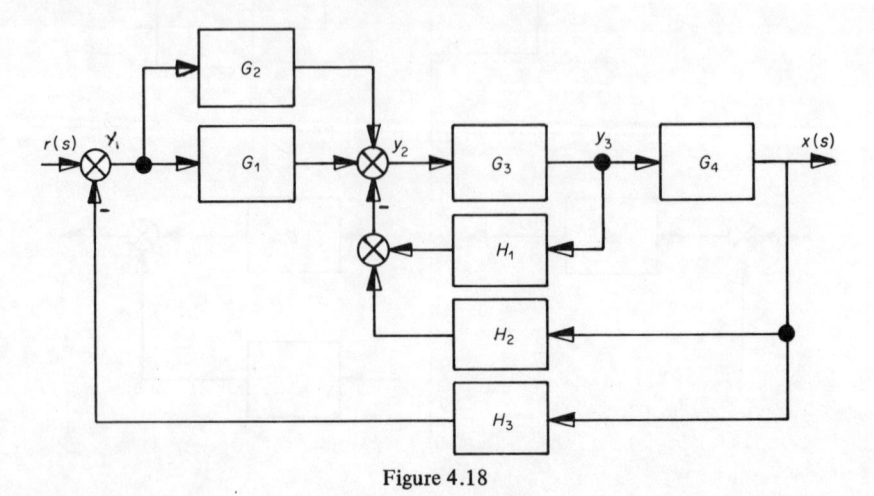

Figure 4.18

Control-system Representation

given in figure 4.17, where the identities are two-way; the following examples will illustrate their use. (Proofs will be left to the reader as an exercise in block-diagram manipulation.)

Example 4.3
Use block-diagram manipulation to obtain the over-all transfer function of the system shown in figure 4.18.

The over-all transfer function can be obtained in three steps, as shown in figure 4.19.

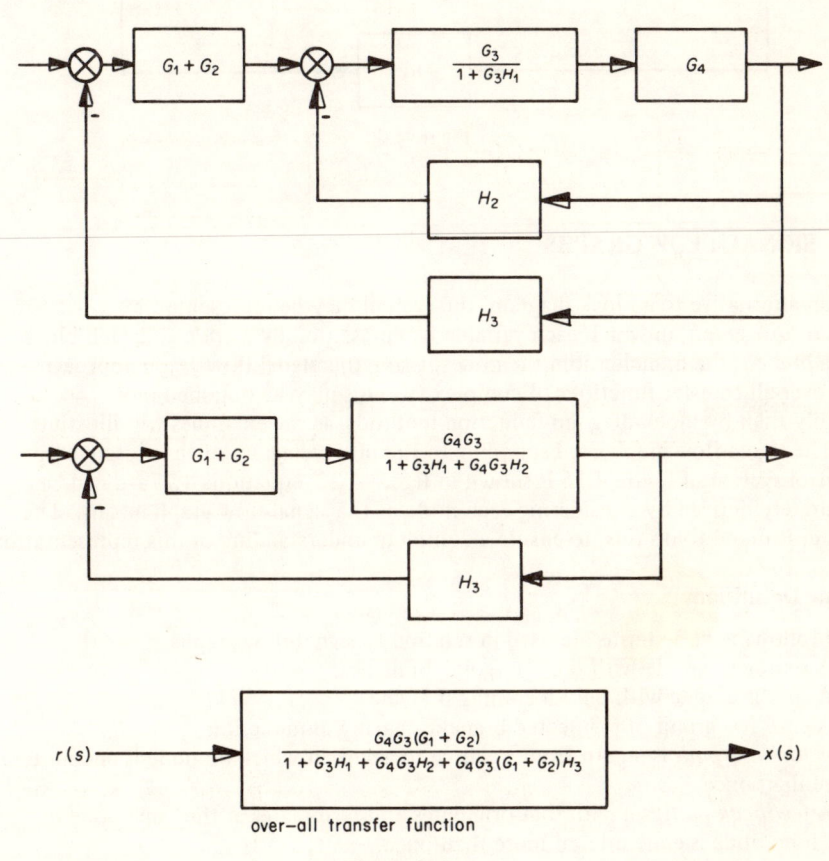

Figure 4.19

Example 4.4
Use block-diagram manipulation to obtain the over-all transfer function of the system shown in figure 4.20.

Four steps are required to obtain the over-all transfer function, as shown in figure 4.21.

Figure 4.20

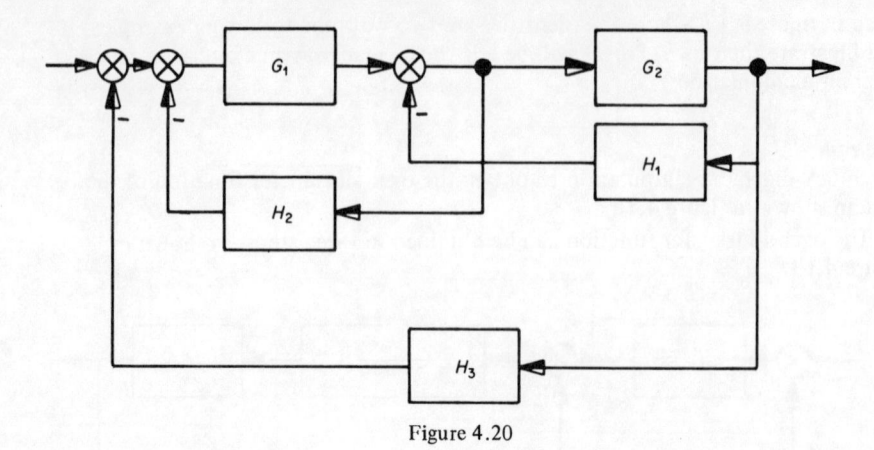

4.4 SIGNAL-FLOW GRAPHS

As an alternative to a block diagram, the system may be represented by a *signal-flow graph*, in which each variable is represented by a *node* and each block by a *branch*; the branches join the nodes. Using this signal-flow-graph approach the over-all transfer functions of complex systems may be obtained more readily than by block-diagram-reduction methods, as the examples will illustrate.

The signal-flow graph corresponding to the block diagram of the closed-loop control system of figure 4.15 is shown in figure 4.22; equations 4.17a–d can be separately derived by considering each node of the signal-flow graph in turn. The reader is urged to do this, to ensure a complete understanding of this representation.

Some Definitions

The following basic terms are used in relation to signal-flow graphs.
A *source* is a node with only outgoing branches.
A *sink* is a node with only incoming branches.
A *path* is a group of connected branches having unidirection.
A *forward path* is a path from a source to a sink in which no node is encountered more than once.
A *feedback path* is a path that originates and terminates at the same node in which no node is encountered more than once.
The *path gain* is the product of the branch gains along the forward path.
The *loop gain* is the product of the branch gains along the feedback path.

Example 4.5
The signal-flow graphs for the block diagrams of figures 4.18 and 4.20 are shown in figures 4.23 and 4.24 respectively.

Control-system Representation

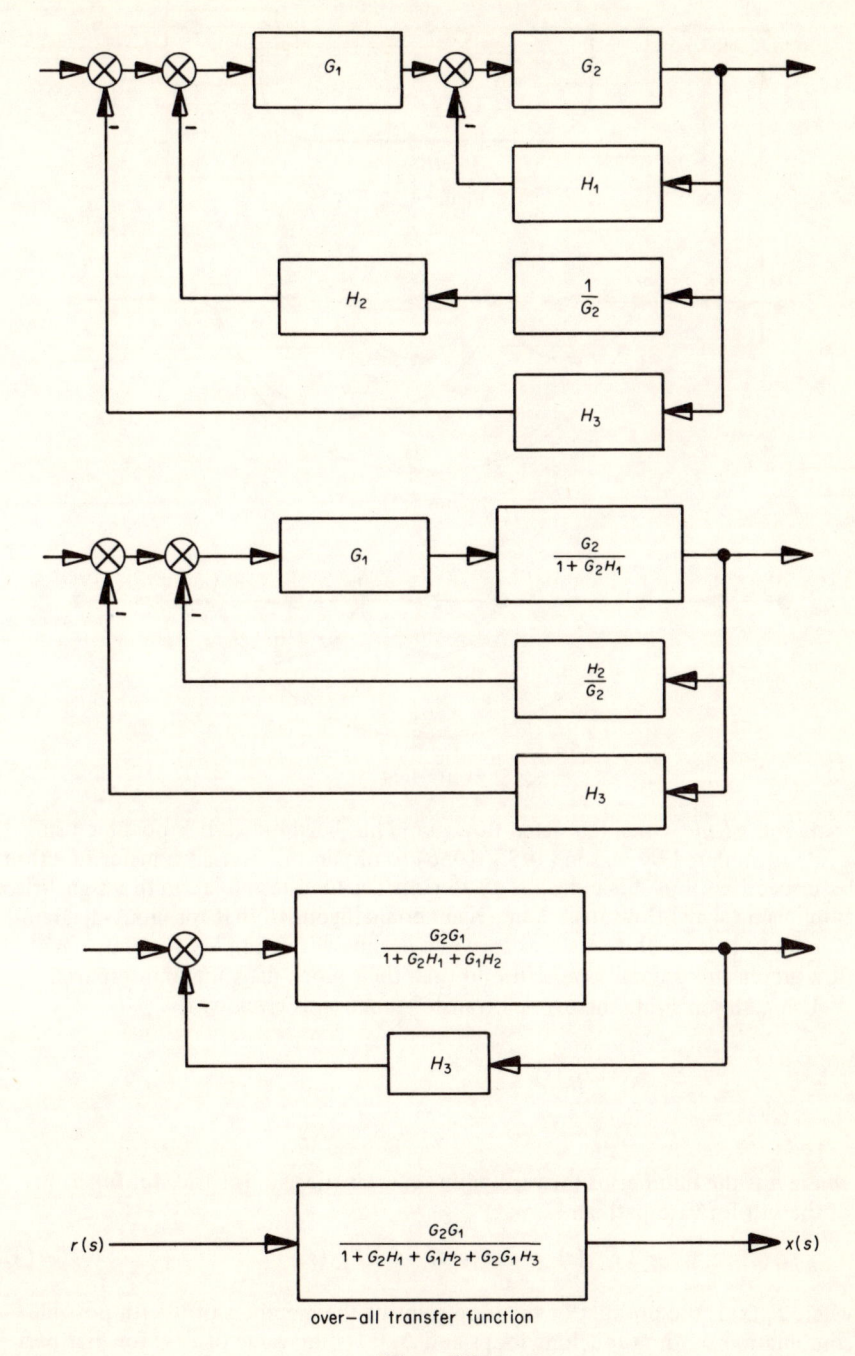

Figure 4.21

Introduction to Control Theory

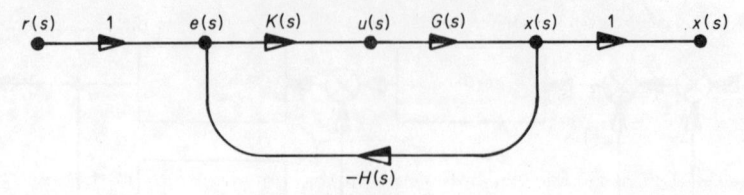

Figure 4.22

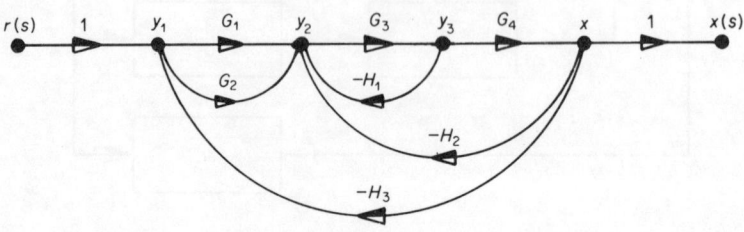

Figure 4.23

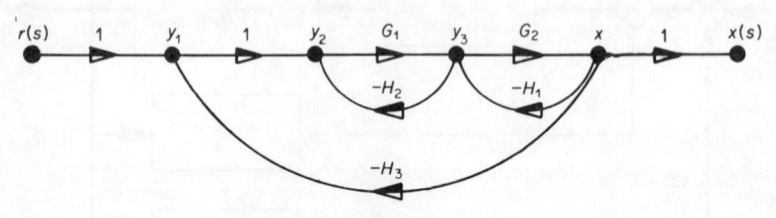

Figure 4.24

Mason's Rule. Once the signal-flow graph has been drawn it is possible using a rule formulated by Mason (1953, 1956) to obtain the over-all transfer function by inspection; thus it is unnecessary to consider the simplification of a signal-flow graph using signal-flow algebra in a manner analogous to that for block-diagram reduction. It is for this reason that, when dealing with complex systems, signal-flow graphs are generally more useful than their block-diagram counterparts.

Using Mason's rule the over-all transfer function is given by

$$G'(s) = \frac{\sum_{k=1}^{n} g_k(s) \Delta_k(s)}{\Delta(s)} \tag{4.25}$$

where n is the number of forward paths, $g_k(s)$ is the gain (or transfer function) of the kth forward path and

$$\Delta(s) = 1 - \sum_i P_{i1}(s) + \sum_i P_{i2}(s) - \sum_i P_{i3}(s) + \ldots \tag{4.26}$$

where $P_{ij}(s)$ is the product of the loop gains of the members of the ith possible combination of j non-touching loops and $\Delta_k(s)$ is the value of $\Delta(s)$ for that part of the graph not touching the kth forward path.

To illustrate the application of the rule, obtain the transfer function of the closed-loop control system shown in figure 4.22. Since there is only one forward path then

$$G'(s) = \frac{g_1(s)\Delta_1(s)}{\Delta(s)}$$

and $g_1(s) = G(s)K(s)$. There is only one non-touching loop; therefore, $i = 1$ and $j = 1$. Hence

$$\Delta(s) = 1 + G(s)K(s)H(s)$$

and

$$\Delta_1(s) = 1$$

Therefore

$$G'(s) = \frac{G(s)K(s)}{1 + G(s)K(s)H(s)}$$

Example 4.6
Obtain the over-all transfer function of the signal-flow graph of figure 4.24 using Mason's rule.

Again there is only one forward path and $g_1(s) = G_2(s)G_1(s)$, while (omitting the s)

$$\Delta(s) = 1 + G_1H_2 + G_2H_1 + G_2G_1H_3$$

since there are three combinations of single loops only, that is, $i = 3, j = 1$. The two inner loops are *not* non-touching since they share a common node. Also $\Delta_1(s) = 1$. Therefore

$$G'(s) = \frac{G_2(s)G_1(s)}{1 + G_1(s)H_2(s) + G_2(s)H_1(s) + G_2(s)G_1(s)H_3(s)}$$

The reader should repeat the exercise for figure 4.23.

Example 4.7
Figure 4.25 shows the block diagram of a control system with input $r(t)$, output $x(t)$ and disturbance $d(t)$. Draw a signal-flow graph for the system and hence obtain

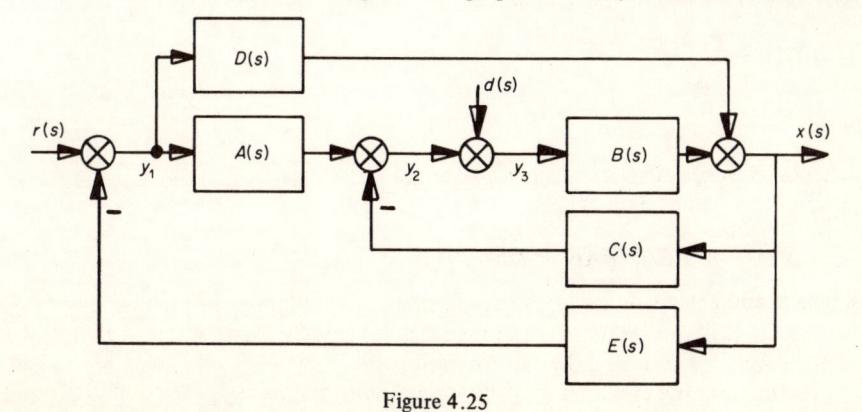

Figure 4.25

the transfer functions relating $x(s)$ to $r(s)$ and $d(s)$. What is the condition that the system shall be stable?

First define the variables y_1, y_2 and y_3, as shown in Fig. 4.25; then the signal-flow graph of figure 4.26 can be drawn.

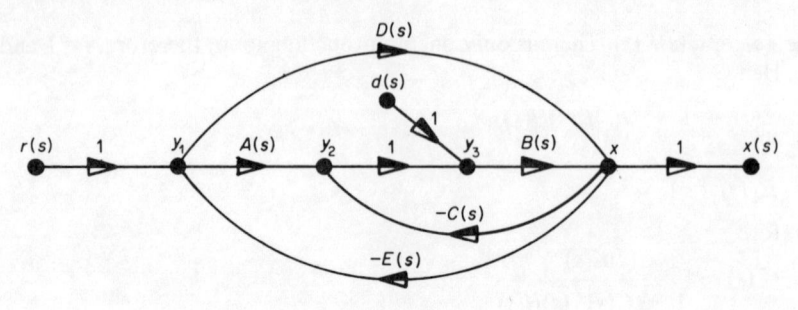

Figure 4.26

Use Mason's rule to obtain initially the transfer function between input $r(s)$ and output $x(s)$, that is, $G_1'(s)$, where

$$G_1' = \frac{g_1 \Delta_1 + g_2 \Delta_2}{\Delta}$$

since there are two forward paths, and

$$\Delta = 1 + BC + BAE + DE$$
$$G_1' = BA; \quad g_2 = D$$
$$\Delta_1 = 1; \quad \Delta_2 = 1$$

Therefore

$$G_1' = \frac{BA + D}{1 + BC + BAE + DE}$$

Similarly obtain the transfer function between disturbance $d(s)$ and output $x(s)$, that is, $G_2'(s)$, where

$$G_2' = \frac{g_1 \Delta_1}{\Delta}$$

In this case

$$g_1 = B$$

while

$$\Delta = 1 + BC + BAE + DE$$

as before and

$$\Delta_1 = 1$$

Therefore

$$G_2' = \frac{B}{1 + BC + BAE + DE}$$

Thus, since the system is linear, it follows from the principle of superposition that the complete transfer function is

$$x(s) = \frac{B(s)A(s) + D(s)}{1 + B(s)C(s) + B(s)A(s)E(s) + D(s)E(s)} r(s)$$

$$+ \frac{B(s)}{1 + B(s)C(s) + B(s)A(s)E(s) + D(s)E(s)} d(s)$$

and stability is guaranteed if the roots of the characteristic equation of the system, $1 + BC + BAE + DE = 0$ (that is, the denominator of the transfer function set equal to zero) have negative real parts.

4.5 CONCLUDING REMARKS

This chapter has been concerned with the various methods of control-system representation—transfer functions, block diagrams and signal-flow graphs. It is now possible to analyse the behaviour of closed-loop control systems and to investigate the effect of certain parameters upon this behaviour; in the next chapter, analysis in the *time domain* is considered.

REFERENCES

Di Stefano, J. J., Stubberud, A. R., and Williams, I. J., *Theory and Problems of Feedback and Control Systems* (McGraw-Hill, New York, 1967).
Mason, S. J., 'Feedback Theory—Some Properties of Signal-flow Graphs', *Proc. I.R.E.*, 41 (1953) 1144-56.
Mason, S. J., 'Feedback Theory—Further Properties of Signal-flow Graphs', *Proc. I.R.E.*, 44 (1956) 920-26.
Melsa, J. L., and Schultz, D. G., *Linear Control Systems* (McGraw-Hill, New York, 1969).
Murphy, G. J., *Basic Automatic Control Theory* (Van Nostrand, New York, 1966).

PROBLEMS

4.1 Name and identify the individual parts of naturally occurring control systems. Draw a block diagram of a gun-fire control system: is it a servomechanism or a regulator? Discuss the driving of a motor car as a closed-loop control system. What part does the driver play and would it be possible to dispense with his services? Develop a possible control system for a windmill such that the vanes are always directed towards the wind.
In each of these systems what is the source of energy?

Introduction to Control Theory

4.2 Obtain the transfer function of the two inverted pendulums shown in figure 4.27. Comment on whether it is always possible to control such a system. What happens when the pendulums are of the same length? Think of other possible inverted-pendulum configurations and develop their transfer functions.

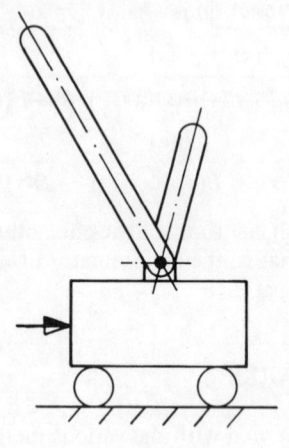

Figure 4.27

4.3 Determine the transfer functions of the simple networks shown in figure 4.28.

4.4 Obtain the transfer functions $\phi(s)/\phi_r(s)$ and $x(s)/x_r(s)$ of the block diagram shown in figure 4.29 and draw the corresponding signal-flow graph. Suggest a system with added closed-loop control that this block diagram represents.

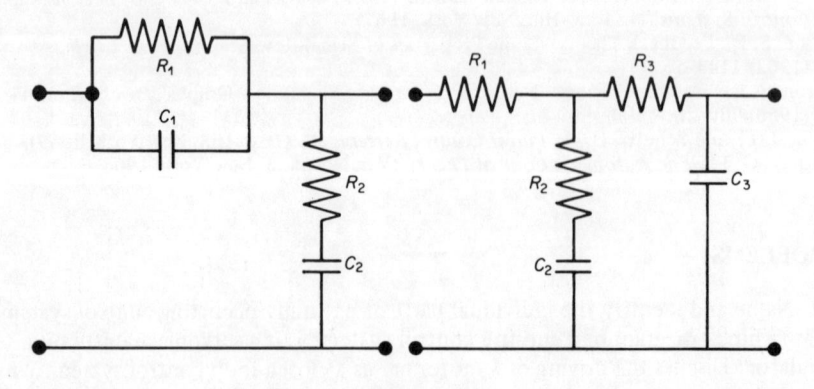

Figure 4.28

4.5 Figure 4.30 shows the block diagram of a control system. Draw a signal-flow graph and thence, or otherwise, obtain the transfer functions relating $x(s)$ to $r(s)$ and $d(s)$. What is the condition that the system should be stable?

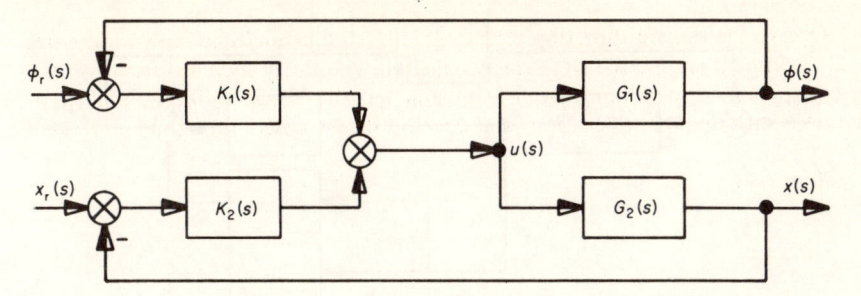

Figure 4.29

4.6 Use Mason's rule to obtain the over-all transfer function of the system shown in figure 4.31.

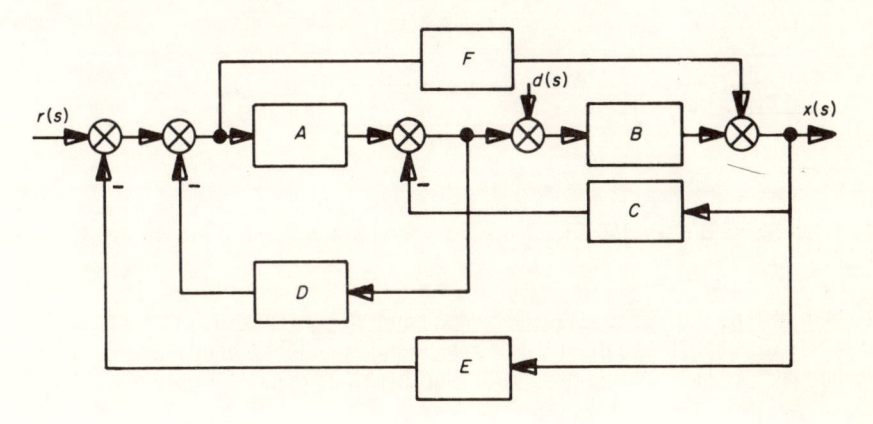

Figure 4.30

4.7 Determine the transfer functions $x_1(s)/r_1(s)$, $x_1(s)/r_2(s)$, $x_2(s)/r_1(s)$ and $x_2(s)/r_2(s)$ of the multi-loop system shown in figure 4.32 using both block-diagram reduction and Mason's rule.

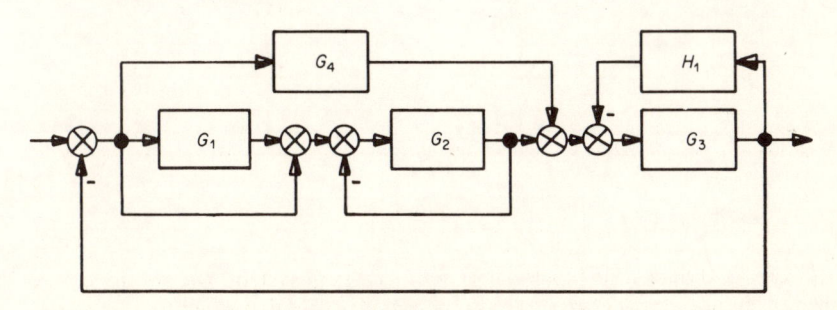

Figure 4.31

Introduction to Control Theory

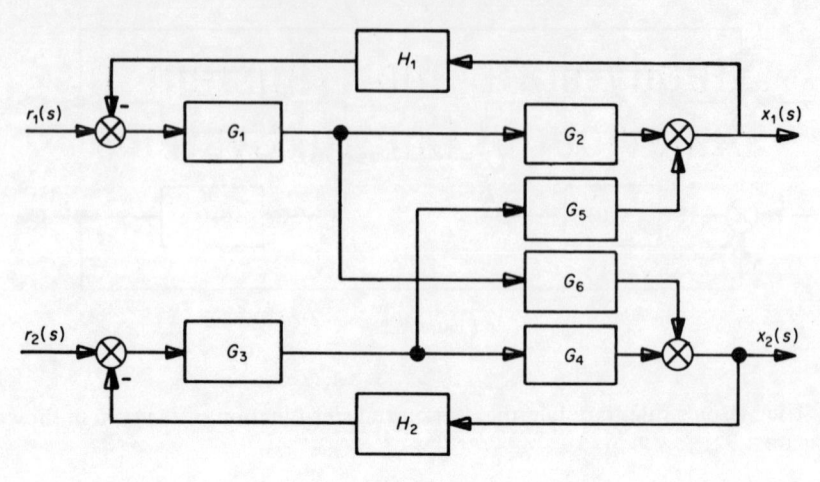

Figure 4.32

5 Steady-state and Transient Behaviour of Control Systems

[Handwritten annotations:]
$x(s) = Kg(s)[r(s) - x(s)]$ $e(s) = r(s) - x(s)$
$x(s)(1 + Kg(s)) = Kg(s) r(s)$ $u(s) = K e(s)$? $d(s)$
$\dfrac{x(s)}{r(s)} = \dfrac{Kg(s)}{1 + Kg(s)}$ $x(s) = g(s) u(s)$

Before embarking on the analysis and design of closed-loop control systems it is essential to have a thorough understanding of the steady-state and transient behaviour of unity-feedback systems, as shown in figure 5.1. Here $G(s)$ is the open-loop transfer function, that is, the transfer function of the plant whose output is to be controlled, and is assumed to be *fixed*; K is taken to be time-independent and is known as the *controller gain*. In chapter 8, controllers of other, more complicated forms are considered.

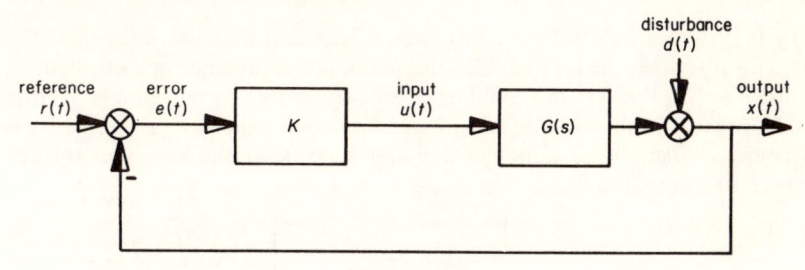

Figure 5.1

This chapter is mainly concerned with the behaviour of closed-loop control systems in the time domain, that is, their transient behaviour (behaviour in the frequency domain is considered in chapter 7), and discusses in detail the effect of controller gain K and some control configurations upon the various performance criteria—including absolute stability, steady-state accuracy, relative stability and transient behaviour, and sensitivity.

Throughout, the treatment relies heavily on results from preceding chapters.

5.1 PERFORMANCE CRITERIA

The purpose of a closed-loop control system is to transmit the reference or input signal $r(t)$ accurately to the output $x(t)$ without allowing excessive error, in the presence of disturbances. Indeed the most desirable response is that which is

Introduction to Control Theory

identical to the input at all times but because of the nature of dynamical systems it is realised that this is not possible. Thus, the four main criteria are as follows.

(1) The system should be absolutely stable: when it is excited, it should settle out to some steady value and not exhibit continuous and growing oscillations. This is shown in figure 5.2 for both step input and *ramp* input, defined as $r(t) = ct$.

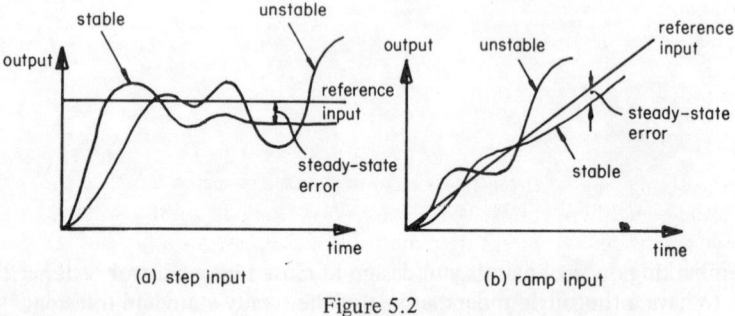

Figure 5.2

(2) The system should be accurate in the steady state; that is, at $t = \infty$, the output $x(t)$ should be equal to the input $r(t)$ or nearly so. In figure 5.2 the stable responses both give a steady-state error, that is, $e_{SS} = r(t) - x(\infty)$, and it is desirable that this error should be minimised; in many cases it can be reduced to zero, and in others its presence is not too important.

(3) The system should exhibit a satisfactory transient response, such that the output follows the input at all times, even in the presence of disturbances; in other words, the system should have good *dynamic* accuracy and be *relatively* stable. Figure 5.3 shows systems with good dynamic accuracy (a) and with poor dynamic accuracy (b, c), although all systems are *stable* and have *zero* steady-state error when following a step change in reference value.

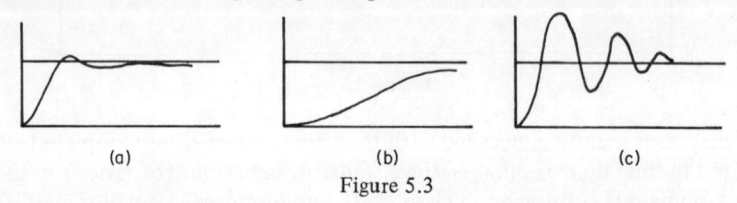

Figure 5.3

(4) The system should be insensitive to changes in system parameters: as plants become older their dynamic behaviour changes—that is, $G(s)$ changes—and a certain degree of insensitivity to these changes is essential to the control system.

These criteria can be interpreted more precisely in engineering terms in many different ways, depending upon the control system being designed—for example, is the system a regulator with disturbances or a servomechanism? Has the system auxiliary feedback loops? What types of input has the system to follow?—and also upon the technique being used, that is, with respect to a response in the *time domain*, or with reference to the roots of the characteristic equation, or with reference to a response in the *frequency domain*. The last two cases are discussed in subsequent **chapters,** whilst the first is considered in this.

5.2 ABSOLUTE STABILITY: THE METHOD OF ROUTH

It has been stated previously that a system is stable if the roots of its characteristic equation (or of the denominator of its transfer function set equal to zero) have negative real parts. Thus, in the case of the control system in figure 5.1, the closed-loop transfer function is

$$G'(s) = \frac{KG(s)}{1 + KG(s)} \tag{5.1}$$

while the roots in question are those of

$$1 + KG(s) = 0 \tag{5.2}$$

and if the system is stable these roots have negative real parts.

Computation of these roots can be extremely time-consuming, as seen previously, but there is an alternative method of assessing whether a system is stable, on the threshold of stability or unstable. This method is due to Routh (1877) and indicates the presence and number of unstable roots in a characteristic equation but *not* their *value*. It proceeds as follows. Form the characteristic equation of the system, that is, $1 + KG(s) = 0$, as

$$a_0 s^n + a_1 s^{n-1} + a_2 s^{n-2} + \ldots + a_{n-2} s^2 + a_{n-1} s + a_n = 0 \tag{5.3}$$

and assume that all coefficients are positive; this is a *necessary* (but not *sufficient*) condition for the system to be stable. Thus, if all coefficients are *not* positive the system is unstable.

From this, form the Routh array of coefficients

$$\begin{array}{llllll} s^n: & a_0 & a_2 & a_4 & a_6 & \ldots \\ s^{n-1}: & a_1 & a_3 & a_5 & a_7 & \ldots \\ s^{n-2}: & b_1 & b_2 & b_3 & \ldots \\ s^{n-3}: & c_1 & c_2 & \ldots \\ s^{n-4}: & d_1 & \ldots \end{array}$$

where

$$b_1 = \frac{a_1 a_2 - a_0 a_3}{a_1}, \quad b_2 = \frac{a_1 a_4 - a_0 a_5}{a_1}, \quad b_3 = \frac{a_1 a_6 - a_0 a_7}{a_1}, \ldots$$

$$c_1 = \frac{b_1 a_3 - a_1 b_2}{b_1}, \quad c_2 = \frac{b_1 a_5 - a_1 b_3}{b_1}, \ldots$$

$$d_1 = \frac{c_1 b_2 - b_1 c_2}{c_1}, \ldots$$

...

When *complete*, the array comprises $n + 1$ rows.

Thus, if the array is complete and if *none* of the elements in the first column vanishes, then a *sufficient* condition for the system to be stable (and hence for the characteristic equation to have roots with negative real parts) is for all these elements to be *positive*. Furthermore, if these elements are not all positive then the number of sign changes in the first column indicates the number of roots with positive real parts and hence system instability.

Example 5.1
Use Routh to indicate whether the characteristic equation

$$s^4 + 2s^3 + 6s^2 + 7s + 5 = 0$$

represents a stable system or not.

Form the Routh array

s^4:   1   6   5

s^3:   2   7

s^2: $\dfrac{(2)(6) - (1)(7)}{2} = 2.5$ $\dfrac{(2)(5) - (1)(0)}{2} = 5$

s: $\dfrac{(2.5)(7) - (2)(5)}{2.5} = 3$

c: 5

This is the complete array of five rows: the first column is *all* positive and thus the system is stable.

Note that it is permissible to *normalise* any row by dividing or multiplying all the elements in that row by a constant factor; in many cases this reduces the amount of arithmetic.

Example 5.2
Use Routh to determine the range of K for which the control system of figure 5.4 is stable.

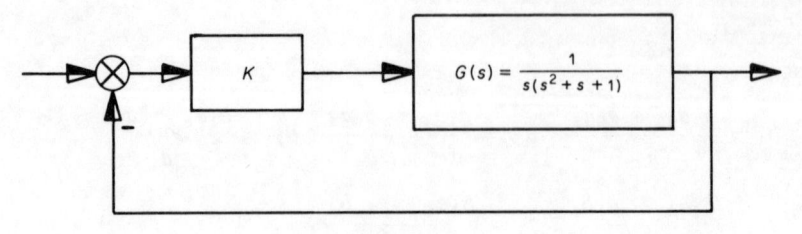

Figure 5.4

The closed-loop transfer function is

$$G'(s) = \frac{KG(s)}{1 + KG(s)} = \frac{K}{s(s^2 + s + 1) + K}$$

Steady-state and Transient Behaviour of Control Systems

and the characteristic equation

$$s^3 + s^2 + s + K = 0$$

determines the system stability. Form the Routh array

$$
\begin{array}{lll}
s^3: & 1 & 1 \\
s^2: & 1 & K \quad \text{(auxiliary equation)} \\
s: & \dfrac{1-K}{1} & \\
c: & K &
\end{array}
$$

which is complete. Thus, for stability, from the third row $1 - K > 0$ and from the fourth row $K > 0$, that is, $0 < K < 1$.

If, for example, $K = 0.5$, then the first column is all positive and hence the system is stable. If $K = 1.5$, the first column is

1
1
−0.5
1

which has *two* sign changes, one from 1 to −0.5 and one from −0.5 to 1. Thus the system is unstable and the characteristic equation has two roots with positive real parts.

If $K = 1$, the s row is completely zero, and it means, in this case, that the characteristic equation has roots on the imaginary axis (that is, roots with *zero* real parts), which can be calculated by forming the auxiliary equation from the s^2 row of the array with $K = 1$, that is

$$1 \times s^2 + 1 = 0$$

Therefore $\quad s^2 = -1$

$$s = \pm j$$

indicating that the system is on the *threshold* of instability and, once excited, will oscillate continuously. The reader should check that, when $K = 1$, $(s^2 + 1)$ is a factor of the characteristic equation.

Presence of a Zero in the First Column. If, when forming the array, the first element in any row is zero whilst the remaining elements in that row are non-zero, it is necessary to replace the first-element zero by an arbitrarily small quantity δ and then to complete the array in the usual manner. The stability is assessed by obtaining the limiting value of the first column as $\delta \to 0$.

Example 5.3
Show that the characteristic equation

$$s^6 + s^5 + 3s^4 + 3s^3 + 2s^2 + s + 1 = 0$$

is unstable and determine the number of roots with positive real parts.

Form the Routh array

s^6: 1 3 2 1
s^5: 1 3 1
s^4: $0 \Rightarrow \delta$ 1   1   (replace 0 by δ)
s^3: $\dfrac{3\delta-1}{\delta}$ $\dfrac{\delta-1}{\delta}$
s^2: $\dfrac{4\delta-1-\delta^2}{3\delta-1}$ 1
s: $\dfrac{\left(\dfrac{4\delta-1-\delta^2}{3\delta-1}\right)\left(\dfrac{\delta-1}{\delta}\right) - \left(\dfrac{3\delta-1}{\delta}\right)}{\left(\dfrac{4\delta-1-\delta^2}{3\delta-1}\right)}$

c: 1

Now, as $\delta \to 0$, the first column will be

positive

positive

positive

negative

positive

negative

positive

and there will be four roots with positive real parts.

Presence of an All-zero Row. If, when forming the array, a complete row of zeros is obtained, it indicates the presence of a pair of roots that are of equal magnitude but of opposite sign, that is, a pair of imaginary roots $\pm ja$ (as in example 5.2 when $K = 1$) or a negative root and positive root of equal magnitude $\pm b$. To continue, it is necessary to *differentiate* the equation formed from the coefficients of the preceding row (with respect to s) and to use the resulting coefficients in place of the all-zero row. This equation will have a pair of roots of equal magnitude and opposite sign and will be identical to the roots of the original equation.

Example 5.4
Use Routh to determine the stability of the system whose characteristic equation is

$$s^3 + 3s^2 + 2s + 6 = 0$$

The Routh array is

s^3:   1   2
s^2:   3   6   (auxiliary equation)
s: 0   0

Form the equation of the preceding row, that is

$$3s^2 + 6 = 0$$

and differentiate to give

$$6s = 0$$

Thus, the new array will be

s^3:   1   2
s^2:   3   6
s: 6   0
c: 1

and since there are no changes in sign the system will *not* be unstable but will have a pair of imaginary roots given by the auxiliary equation

$$3s^2 + 6 = 0$$

Therefore

$$s = \pm j\sqrt{2}$$

The system will be on the threshold of instability. To calculate the third root of the characteristic equation divide out the factor $s^2 + 2$, as follows

$$\begin{array}{r}
s + 3 \\
s^2 + 0s + 2 \overline{\smash{\big)}\, s^3 + 3s^2 + 2s + 6} \\
\underline{s^3 + 0s^2 + 2s} \\
3s^2 + 0s + 6 \\
\underline{3s^2 + 0s + 6} \\
\end{array}$$

that is, $(s + 3)$ is a factor and the third root is at -3.

5.3 STEADY-STATE ANALYSIS AND CLASSIFICATION OF SYSTEMS

Once the absolute stability of a closed-loop control system has been investigated to ensure that the system possesses a steady state the next step is to analyse its steady-state performance or static accuracy, since it is desirable that the ultimate response of the system should be equal to the reference or input signal. Thus, the error in steady state is a measure of the system's steady-state performance.

This error will depend upon the nature of the open-loop transfer function $G(s)$ and the type of input signal that the control system has to follow and also upon the effect of disturbance signals. To avoid investigating the numerous variations that can exist, it is possible to classify the steady-state behaviour depending upon the system type number (to be defined later), which leads to a definition of error constants.

Consider now the unity-feedback system of figure 5.5, where $G(s)$ is the open-loop transfer function and K is the gain of the controller. It is assumed that other forms of control system are put into the form of figure 5.5 using the block-diagram-reduction methods of chapter 4 before continuing with the steady-state analysis.

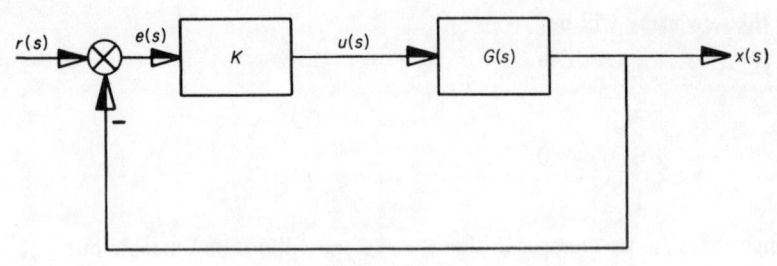

Figure 5.5

Then, the transform of the error $e(s)$, between the reference signal and the output, is given by

$$e(s) = \frac{1}{1 + KG(s)} r(s) \qquad (5.4)$$

The ultimate or steady-state error of the system is defined by e_{ss}, where

$$e_{ss} = \lim_{t \to \infty} e(t) \qquad (5.5)$$

and therefore by the final-value theorem

$$e_{ss} = \lim_{s \to 0} se(s) \qquad (5.6)$$

that is

$$e_{ss} = \lim_{s \to 0} \left[s \frac{1}{1 + KG(s)} r(s) \right] \qquad (5.7)$$

This steady-state error e_{ss} should be ideally be zero for a large class of input signals, including step inputs, ramp inputs and acceleration inputs. In practice, however, the design will usually be less restrictive since, for example, if it is expected that the system will be required to follow step changes and slowly varying changes, the need for zero steady state error when subjected to acceleration inputs may be relaxed. From equation 5.7 this error will also depend upon the form of $G(s)$.

Classification of Systems

The *forward transference* $KG(s)$ of the control system in figure 5.5 will in general be expressed as

$$KG(s) = K \frac{K_G(s + a_1)(s + a_2) \ldots (s^2 + bs + c) \ldots}{s^l(s + d_1)(s + d_2) \ldots (s^2 + es + f) \ldots} \quad (5.8)$$

$$= \frac{K \sum_{k=0}^{m} A_k s^k}{s^l \sum_{k=0}^{n-l} B_k s^k}, \quad n \geqslant m + 1 \quad (5.9)$$

where the *order* of the system is defined as the highest power of s in the denominator—that is, n; the *rank* of the system is defined as the difference between the highest power of s in the denominator and that in the numerator—that is, $n - m \geqslant 1$; and the *class* (or *type number*) of the system is defined as the power of the factor s in the denominator—that is, l. The class is equal to the number of open-loop *integrators* in cascade from $e(t)$ to $x(t)$ since

$$\mathcal{L}\left[\int_0^t f(\tau)d\tau\right] = \frac{1}{s} f(s) \quad (5.10)$$

from equation 3.30.

Example 5.5
State the order, rank and type number of the systems with open-loop transfer function

(i) $\quad G(s) = \dfrac{s + 2}{s^4 + 3s^3 + 3s^2 + s}$

(ii) $\quad G(s) = \dfrac{1}{s^3 (s + 2)(s + 1)}$

(iii) $\quad G(s) = \dfrac{s^2 + s + 1}{(s + 2)(s^2 + s + 4)}$

(i) In this case the order is 4, the rank is $4 - 1 = 3$ and the type number is 1 since the denominator can be written as $s(s^3 + 3s^2 + 2s + 1)$.

(ii) Here the order is 5, the rank is 5 and the type number is 3 since s^3 is a factor.

(iii) Finally the order is 3, the rank is 1, and the type number is 0.

Steady-state Error due to Step Input

In this case for the unit step input

$$r(s) = \frac{1}{s} \quad (5.11)$$

and

$$e_{ss} = \lim_{s \to 0} \left[\frac{1}{1 + KG(s)} \right] \qquad (5.12)$$

$$= \frac{1}{1 + \lim_{s \to 0} [KG(s)]} \qquad (5.13)$$

$$= \frac{1}{1 + K_p} \qquad (5.14)$$

where

$$K_p = \lim_{s \to 0} [KG(s)]$$

and is known as the *position error constant* since the input is a step.

For a type-0 system (that is, when $l = 0$ in equation 5.9)

$$K_p = K \frac{A_0}{B_0} = KC_0 \qquad (5.15)$$

and

$$e_{ss} = \frac{1}{1 + KC_0}$$

which is non-zero and decreases as K is increased. (Later in this chapter it will become evident that K cannot be increased indefinitely because of stability requirements: hence a compromise must be achieved.)

For a type-1 system, that is, when $l = 1$, $K\hat{p} = \infty$ as a result of the presence of the factor s in the denominator of $KG(s)$ and therefore

$$e_{ss} = 0 \qquad (5.16)$$

indicating ultimate equality between input and output. Thus, if a closed-loop control system is to have zero steady-state error when subjected to a step change, its forward transference $KG(s)$ must be at least of type 1. This is an important result, which will be referred to again later and used in design.

Steady-state Error due to Velocity Input

The Laplace transform of the unit velocity input $r(t) = t$ is

$$r(s) = \frac{1}{s^2} \qquad (5.17)$$

and correspondingly, from equation 5.7, the steady-state error is

Steady-state and Transient Behaviour of Control Systems

$$e_{ss} = \lim_{s \to 0} \left[\frac{1}{s + sKG(s)} \right] \tag{5.18}$$

$$= \frac{1}{\lim_{s \to 0} [sKG(s)]} = \frac{1}{K_v} \tag{5.19}$$

where

$$K_v = \lim_{s \to 0} [sKG(s)]$$

and is known as the *velocity error constant*.

For a type-0 system

$$K_v = 0 \tag{5.20}$$

giving $e_{ss} = \infty$ and indicating that a type-0 system is not capable of following a velocity input.

For a type-1 system

$$K_v = K \frac{A_0}{B_0} = KC_0 \tag{5.21}$$

giving $e_{ss} = 1/KC_0$ and indicating a finite following error once steady conditions have been reached.

For a type-2 system

$$K_v = \infty; \quad e_{ss} = 0 \tag{5.22}$$

indicating a zero steady-state following error.

Steady-state Error due to Acceleration Inputs

Since $r(t) = t^2/2$, say, then $r(s) = 1/s^3$ and the steady-state error is

$$e_{ss} = \lim_{s \to 0} \left[\frac{1}{s^2 + s^2 KG(s)} \right] \tag{5.23}$$

$$= \frac{1}{\lim_{s \to 0} [s^2 KG(s)]} = \frac{1}{K_a} \tag{5.24}$$

where

$$K_a = \lim_{s \to 0} [s^2 KG(s)]$$

is the *acceleration error constant*.

The reader should verify the following results: for a type-0 system

$$K_a = 0$$

giving

$$e_{ss} = \infty$$

for a type-1 system

$$K_a = 0$$

giving

$$e_{ss} = \infty$$

for a type-2 system

$$K_a = KC_0$$

giving

$$e_{ss} = 1/KC_0$$

for a type-3 system

$$K_a = \infty$$

giving

$$e_{ss} = 0$$

The above results (summarised in table 5.1) are for unit input quantities only (for non-unity inputs the results should be multiplied by the appropriate constants); they are shown graphically in figures 5.6 and 5.7 for step and ramp inputs, respectively.

Table 5.1

Type of System	Steady-state Error Constants			Steady-state Errors		
	unit step	unit ramp	unit acceleration	unit step	unit ramp	unit acceleration
0	K_p	0	0	$\dfrac{1}{1+K_p}$	∞	∞
1	∞	K_v	0	0	$\dfrac{1}{K_v}$	∞
2	∞	∞	K_a	0	0	$\dfrac{1}{K_a}$
3	∞	∞	∞	0	0	0

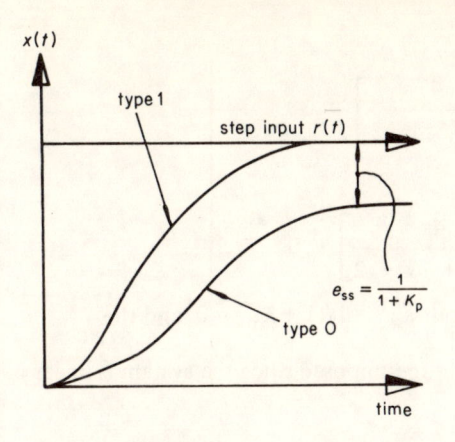

Steady-state and Transient Behaviour of Control Systems 85

Figure 5.6

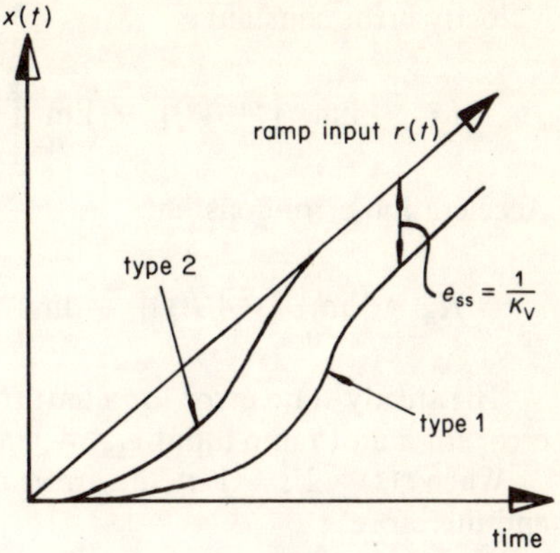

Figure 5.7

Example 5.6
Find the position, velocity and acceleration error constants of the control system shown in figure 5.8 for $K = 4$. Hence deduce the steady-state errors due to a unit step input and a unit ramp input. What is the steady-state error due to the transformed input $r(s) = 2/s - 1/s^2$?

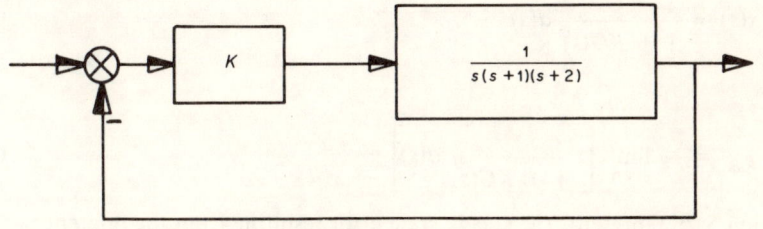

Figure 5.8

First establish that the control system is stable when $K = 4$. Thus

$$G'(s) = \frac{KG(s)}{1 + KG(s)} = \frac{K}{s^3 + 3s^2 + 2s + K}$$

and applying Routh to the characteristic equation $(s^3 + 3s^2 + 2s + 4)$ gives

s^3:	1	2
s^2:	3	4
s:	2	
c:	4	

Since the first column is all positive, the system is stable.
  Position error constant

$$K_p = \lim_{s \to 0} [4G(s)] = \lim_{s \to 0} \left[\frac{4}{s^3 + 3s^2 + 2s} \right] = \infty$$

Velocity error constant

$$K_v = \lim_{s \to 0} [s 4 G(s)] = \lim_{s \to 0} \left[\frac{4}{s^2 + 3s + 2} \right] = 2$$

Acceleration error constant

$$K_a = \lim_{s \to 0} [s^2 4 G(s)] = \lim_{s \to 0} \left[\frac{4s}{s^2 + 3s + 2} \right] = 0$$

The steady-state error for a unit step input $e_{ss}^{\;1} = 1/(1 + K_p) = 0$, and the error for a unit ramp input $e_{ss}^{\;2} = 1/K_v = 1/2$.

When $r(s) = 2/s - 1/s^2$ the errors can be superimposed since the system is linear and therefore

$$e_{ss}^{\;total} = 2 e_{ss}^{\;1} - e_{ss}^{\;2} = -\tfrac{1}{2}$$

Steady-state Error due to Disturbances

Consider an additive disturbance $d(s)$ at the output of the control system; assuming the reference signal to be zero, then

$$x(s) = \frac{1}{1 + KG(s)} d(s) \tag{5.25}$$

and

$$e_{ss} = -\lim_{s \to 0} \left[s \frac{1}{1 + KG(s)} d(s) \right] \tag{5.26}$$

Equation 5.26 represents the steady-state error resulting from the reference signal; it is identical to equation 5.7, except for the sign, and may therefore be analysed in a similar manner. Hence the previous results are directly applicable.

For a disturbance applied to the input of the open-loop plant—that is, between K and $G(s)$—the corresponding results are

$$x(s) = \frac{G(s)}{1 + KG(s)} d(s) \tag{5.27}$$

and

$$e_{ss} = -\lim_{s \to 0} \left[\frac{sG(s)}{1 + KG(s)} d(s) \right] \tag{5.28}$$

In the case of a unit step disturbance

$$e_{ss} = -\frac{1}{\lim_{s \to 0} \left[\dfrac{1}{G(s)} + K \right]} \tag{5.29}$$

5.4 TRANSIENT BEHAVIOUR OF CONTROL SYSTEMS

Before embarking upon the interpretation of system requirements with regard to the transient behaviour of closed-loop control systems following a step change in reference signal, consider the typical behaviour of first- and second-order systems.

First-order Systems

Let the system to be controlled be of first order and represented in transfer-function form by

$$G(s) = \frac{1}{1 + Ts} \tag{5.30}$$

This system is called a *first-order lag* with a *time constant* of T and if excited in *open-loop* by a unit step input its transient reponse is as in figure 3.1, repeated here (figure 5.9b) together with the open-loop block diagram (figure 5.9a).

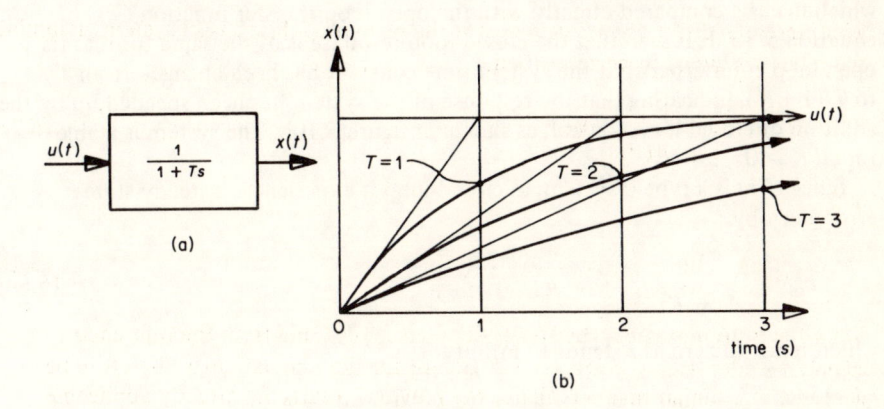

Figure 5.9

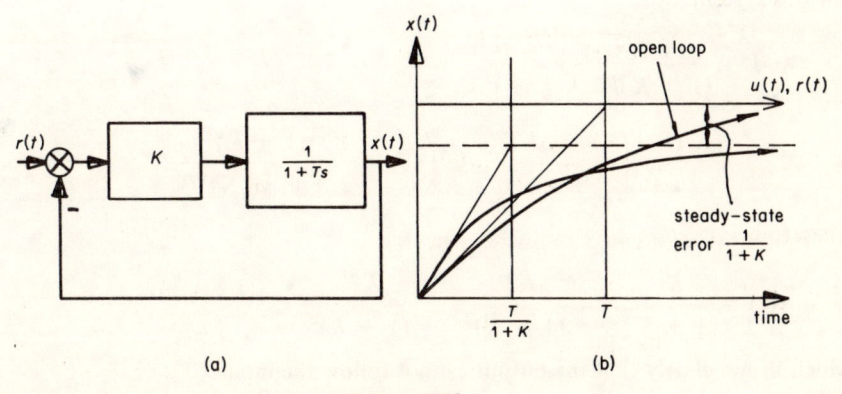

Figure 5.10

Note that the response reaches 66 per cent of its final value after T s.

The first-order lag in *closed-loop* is shown in figure 5.10a, from which the input–output relationship may be directly obtained as

$$x(s) = \frac{K/(1 + Ts)}{1 + K/(1 + Ts)} r(s) \tag{5.31}$$

that is

$$x(s) = \frac{K}{1 + K + Ts} r(s) \tag{5.32}$$

and the closed-loop transfer function is

$$G'(s) = \frac{K/(1 + K)}{1 + [T/(1 + K)]s} \tag{5.33}$$

which may be compared directly with the open-loop transfer function $G(s)$ of equation 5.30. It is seen that the closed-loop response is of the same form as its open-loop counterpart but the system time constant has been changed from T to $T/(1 + K)$, indicating that the response of the system has been speeded up by the addition of closed-loop control, as shown in figure 5.10b. The system is stable for *all* $K > 0$.

Since $G(s)$ is a type-0 system, in closed-loop it experiences a steady-state error given by

$$e_{ss} = \frac{1}{1 + K} \tag{5.34}$$

which tends to zero as K tends to infinity.

Example 5.7
Verify that the system of figure 5.10a cannot follow a velocity input.
For this system

$$x(s) = \frac{K/T}{(1 + K)/T + s} \cdot \frac{1}{s^2}$$

$$= \frac{K/(1 + K)}{s^2} - \frac{KT/(1 + K)^2}{s} + \frac{KT/(1 + K)^2}{s + (1 + K)/T}$$

Therefore

$$x(t) = \frac{K}{1 + K} t - \frac{KT}{(1 + K)^2} + \frac{KT}{(1 + K)^2} e^{-[(1 + K)/T]t}$$

which shows clearly that the output cannot follow the input.

Steady-state and Transient Behaviour of Control Systems

Second-order Systems

In this case the open-loop system is represented by

$$G(s) = \frac{\omega_n^2}{s^2 + 2\xi\omega_n s + \omega_n^2} \tag{5.35}$$

where ξ is the damping factor and ω_n is the undamped natural frequency, and its open-loop step response is shown in detail in figures 3.3 and 3.5 for the cases $\xi > 1$ (overdamped), $\xi = 1$ (critically damped) and $\xi < 1$ (underdamped). The reader is advised to refer back to these figures and refresh his memory with these important responses.

Note again that it is of the utmost importance to understand fully the behaviour of second-order systems such as equation 5.35 because, under certain conditions, such systems can approximate the behaviour of higher-order complex systems. In the next section, for the same reason, system-performance requirements are expressed in terms of the second-order system.

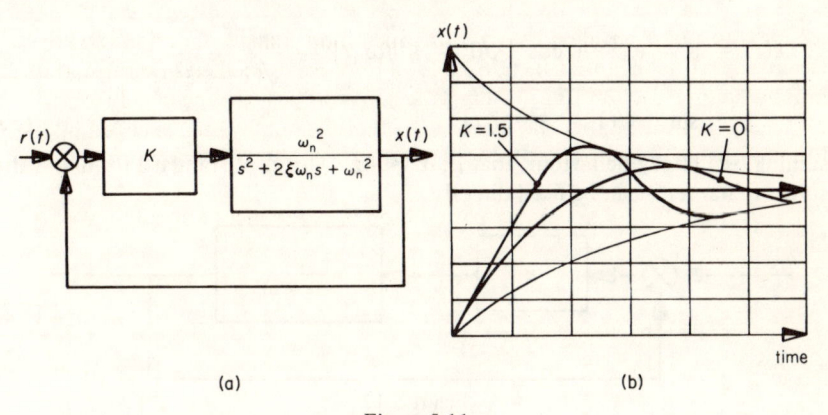

Figure 5.11

The closed-loop transfer function of the system of figure 5.11a is given by

$$G'(s) = \frac{K\omega_n^2/(s^2 + 2\xi\omega_n s + \omega_n^2)}{1 + K\omega_n^2/(s^2 + 2\xi\omega_n s + \omega_n^2)} \tag{5.36}$$

$$= \frac{K\omega_n^2}{s^2 + 2\xi\omega_n s + (1 + K)\omega_n^2} \tag{5.37}$$

which indicates that as K increases so does the *oscillatory nature* of the system, as shown in figure 5.11b with $\xi = 0.5$ and $\omega_n = 1$. Also it can be shown that the system is underdamped or oscillatory for $0 \leq \xi^2 < 1 + K$ rather than $0 \leq \xi < 1$, is always stable, and has a steady-state error

$$e_{ss} = \frac{1}{(1 + K)} \tag{5.38}$$

Performance Criteria of a Second-order System

In order to characterise the behaviour of a second-order system, consider that shown in figure 5.12, where $G(s)$ is of type 1 and hence there is no steady-state error in this case. The closed-loop transfer function is

$$G'(s) = \frac{K}{s^2 + \alpha s + K} \tag{5.39}$$

$$= \frac{\omega_n^2}{s^2 + 2\xi\omega_n s + \omega_n^2} \tag{5.40}$$

where

$$\omega_n = \sqrt{K}; \quad \xi = \frac{\alpha}{2\sqrt{K}}$$

and the step response is

$$x(t) = 1 - e^{-\xi\omega_n t}\cos[\sqrt{(1-\xi^2)}\omega_n t] - \frac{\xi}{\sqrt{(1-\xi^2)}} \cdot \sin[\sqrt{(1-\xi^2)}\omega_n t] \tag{5.41}$$

assuming oscillatory behaviour, that is, $\alpha^2 < 4K$. This is of standard form and the results of chapter 3 can be used directly.

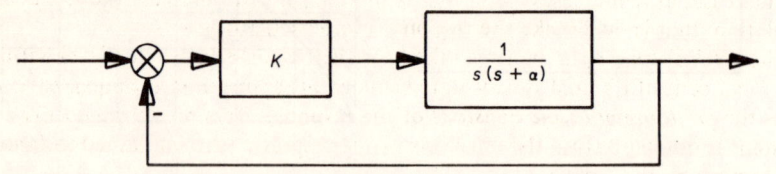

Figure 5.12

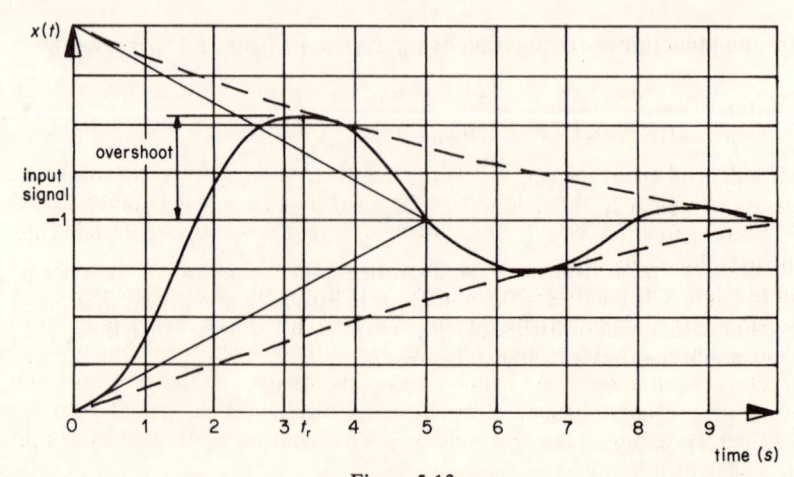

Figure 5.13

Steady-state and Transient Behaviour of Control Systems

Referring to figure 5.13, which shows the step response of figure 5.11 with $K = 1$ and $\alpha = 0.4$, the following definitions can be made. (Note that it is usual to characterise the transient behaviour in terms of a step response.)

The *rise time* is the time taken for the response to reach its first or peak overshoot

$$t_r = \frac{\pi}{\omega_n \sqrt{(1 - \xi^2)}} = \frac{\pi}{\sqrt{(K - \alpha^2/4)}} \tag{5.42}$$

and is seen to decrease as K increases. Thus, increasing the gain will decrease the rise time and hence increase the speed of response.

The *percentage overshoot* is defined as follows

$$\text{Percentage overshoot} = 100 \left(\frac{\text{maximum value of } x(t) - \text{steady-state value}}{\text{steady-state value}} \right) \tag{5.43}$$

$$= 100 \, e^{-\xi\pi/\sqrt{(1-\xi^2)}} = 100 \, e^{-\alpha\pi/\sqrt{(4K - \alpha^2)}} \tag{5.44}$$

It increases as K increases. Thus, increasing the gain will increase the percentage overshoot.

The *frequency of oscillation (natural frequency)* ω_r is given by

$$\omega_r = \omega_n \sqrt{(1 - \xi^2)} = \sqrt{(K - \alpha^2/4)} \tag{5.45}$$

and increases as K increases. Thus increasing the gain will increase the frequency of oscillation, that is, will make the response more oscillatory.

The *settling time* t_s is the time taken for the response to reach a level within 2 to 5 per cent of its final steady-state value and is approximately equal to four times the *predominant time constant* of the response. This predominant time constant is obtained from the envelope of the response (shown dashed in figure 5.13), given by the equation

$$e^{-\xi\omega_n t} \tag{5.46}$$

Hence the predominant time constant is

$$\frac{1}{\xi\omega_n} = \frac{1}{(\alpha/2\sqrt{K})\sqrt{K}} = \frac{2}{\alpha} \tag{5.47}$$

and is independent of the gain K. Thus, to influence this time constant it is necessary to change α, the damping in the plant but, since it is assumed that the plant transfer function $G(s)$ is fixed, alternative methods must be found. This problem is discussed later.

In the design of a closed-loop control system, the speed of response should be reasonably fast, the percentage overshoot not too great and the settling time not too long. However, it is evident from the previous considerations that these are conflicting requirements, and hence the system design must be something of a compromise. Usually, if speed of response and peak overshoot are of prime importance then the damping factor should be $\xi = 0.4$ whilst if settling time is important then a value of 0.7 should be chosen.

Thus, in summary, the damping factor should be selected such that $0.4 \leq \xi \leq 0.7$.

Example 5.8
Given the system of figure 5.12 with $\alpha = 2$, choose K such that the rise time is 1.57 s and calculate the percentage overshoot, the frequency of oscillation and the predominant time constant.

First of all, since the rise time

$$t_r = 1.57 = \frac{\pi}{\sqrt{(K-1)}}$$

the required value of K is 5. Substituting this value of K into the expression for the percentage overshoot gives

$$\text{percentage overshoot} = 100e^{-\alpha\pi/\sqrt{(4K-\alpha^2)}} = 100e^{-2\pi/4} = 21\%$$

Similarly the frequency of oscillation is

$$\omega_r = \sqrt{(K - \alpha^2/4)} = 2 \text{ rad/s}$$

and the predominant time constant is

$$\frac{1}{\xi\omega_n} = 1 \text{ s}$$

The reader should plot the resulting step response.

Addition of Velocity Feedback

To change the predominant time constant of the second-order type-1 system of figure 5.12 it is necessary to modify the amount of damping in the open-loop system, that is, the coefficient α. One way of achieving this is by the addition of *velocity feedback* (see figure 5.14): this involves the addition to the control system of a subsidiary loop that feeds back a signal proportional to the velocity of the output; this signal is only non-zero when the output is changing. Methods of implementation are discussed in a later chapter.

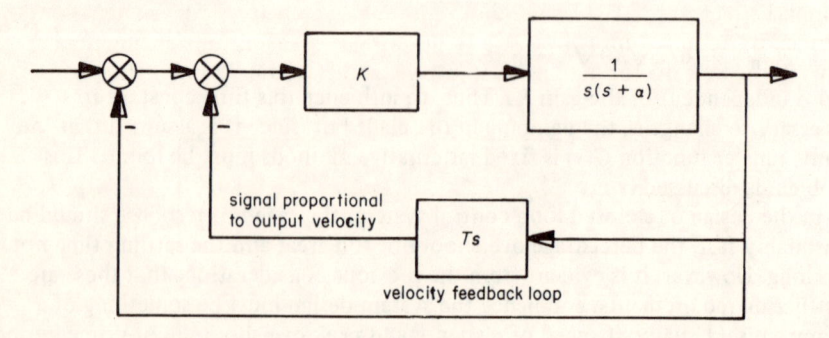

Figure 5.14

The closed-loop transfer function of figure 5.14 is given by

$$G'(s) = \frac{K}{s^2 + (\alpha + KT)s + K} \qquad (5.48)$$

This expression resembles equation 5.39 but α has been replaced by $(\alpha + KT)$. Thus, assuming an oscillatory system, the predominant time constant is given by

$$\frac{2}{\alpha + KT} \qquad (5.49)$$

and is shown in figure 5.15. Hence addition of velocity feedback will have a stabilising effect upon a closed-loop control system, thereby allowing the designer to choose a higher value of the gain K than previously. The reader is urged to confirm this by calculating the various performance indicators and obtaining the transient response of a system with velocity feedback.

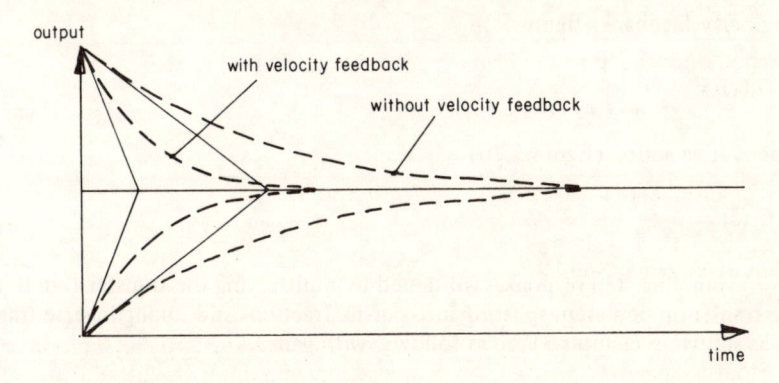

Figure 5.15

Addition of Derivative Action

A similar effect to that achieved by the addition of velocity feedback may be obtained by adding a derivative-type term to the controller transfer function, as shown in figure 5.16 (in this case T is known as the derivative time). The closed-loop

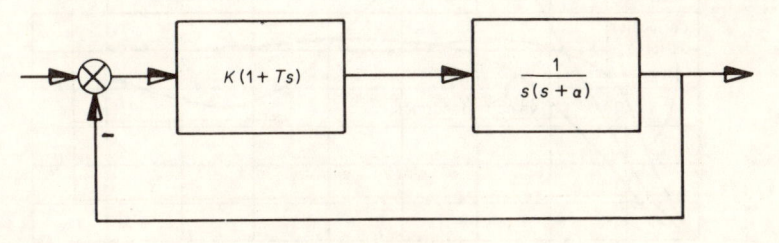

Figure 5.16

transfer function then becomes

$$G'(s) = \frac{K(1 + Ts)}{s^2 + (\alpha + KT)s + K} \tag{5.50}$$

which is identical to the transfer function of the system with velocity feedback—equation 5.48—except for the $(1 + Ts)$ term in the numerator. Thus, addition of derivative action has a similar stabilising effect.

Example 5.9
Compare the step responses of figures 5.12, 5.14 and 5.16 when $\alpha = 0.4$, $K = 1$ and $T = 1$.

The closed-loop transfer functions are as follows: with gain K only (figure 5.12)

$$G'(s) = \frac{1}{s^2 + 0.4s + 1}$$

with velocity feedback (figure 5.14)

$$G'(s) = \frac{1}{s^2 + 1.4s + 1}$$

with derivative action (figure 5.16)

$$G'(s) = \frac{1 + s}{s^2 + 1.4s + 1}$$

The corresponding step responses (obtained by multiplying the transfer functions by the transform of a step, splitting into partial fractions and taking inverse transforms as shown in chapter 3) are as follows: with gain K

$$x(t) = 1 - e^{-0.2t} \cos\sqrt{(0.96)}t - e^{-0.2t} \frac{0.2}{\sqrt{0.96}} \sin\sqrt{(0.96)}t$$

with velocity feedback

$$x(t) = 1 - e^{-0.7t} \cos\sqrt{(0.51)}t - e^{-0.7t} \frac{0.7}{\sqrt{0.51}} \sin\sqrt{(0.51)}t$$

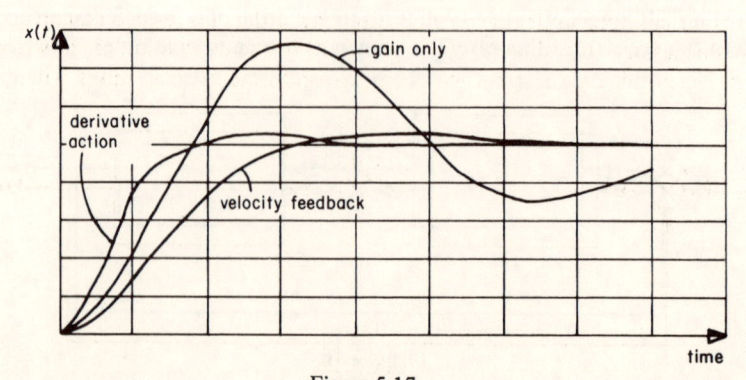

Figure 5.17

Steady-state and Transient Behaviour of Control Systems

with derivative action

$$x(t) = 1 - e^{-0.7t} \cos \sqrt{(0.51)}t + e^{-0.7t} \cdot \frac{0.3}{\sqrt{0.51}} \sin \sqrt{(0.51)}t$$

These responses are shown in figure 5.17. Thus, it is seen that the addition of velocity feedback stabilises the original system whilst the addition of derivative action both stabilises the system and increases the speed of response.

Effect of Disturbances

The output of the system shown in figure 5.18 is given by

$$x(s) = \frac{K}{s(s + \alpha) + K} r(s) + \frac{1}{s(s + \alpha) + K} d(s) \tag{5.51}$$

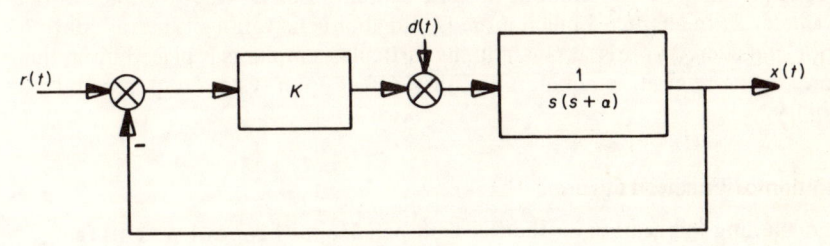

Figure 5.18

Assuming a zero reference input

$$x(s) = \frac{1}{s(s + \alpha) + K} d(s) \tag{5.52}$$

If the disturbance is a unit step change, the resulting steady-state value of the output is

$$x(\infty) = x_{ss} = \frac{1}{K} \tag{5.53}$$

and its transient behaviour is typically of second order. To reduce this error to zero it is necessary to add an integral-type term to the controller, as shown in figure 5.19 (in this case T is known as the integral time or reset time). Then,

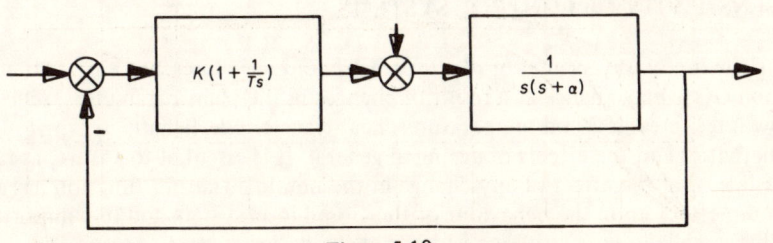

Figure 5.19

assuming a zero reference input, the output is

$$x(s) = \frac{Ts}{Ts^2(s + \alpha) + K(1 + Ts)} d(s) \tag{5.54}$$

corresponding to a *zero* steady-state value for the output, which is correct.

The inclusion of an integral term in the controller reduces to zero the steady-state output resulting from the action of a disturbance and also, in the case of a type-0 system, reduces to zero the steady-state error caused by a finite reference signal, that is, it makes a type-0 system behave like a type-1 system in steady state. This correcting action arises because the integrator output will continually change as long as there is an error present. Once zero error has been achieved, the output from the integrator will remain constant thereby ensuring that the control-system output remains at a value equal to the reference signal.

The disadvantage of integral action is that it tends to have a destabilising effect on the control system, but this can be minimised by careful choice of the parameter T. In chapter 8 much more is said about the various types of control action considered in this section and, in particular, emphasis is placed upon the choice of the relevant parameters and their effect upon system response and stability.

Addition of Further Lags

In concluding this section on the transient behaviour of control systems (and, in particular, second-order systems), it is appropriate to point out that, as the open loop system increases in complexity—that is, as further lags are added, as shown in figure 5.20—the tendency for the closed-loop system to become unstable is increased and the problem of designing a suitable control system becomes more difficult.

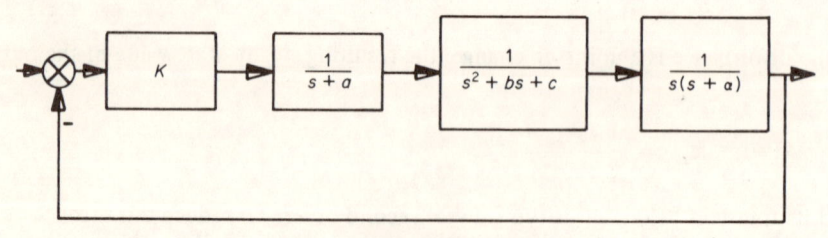

Figure 5.20

5.5 SENSITIVITY OF CONTROL SYSTEMS

During the life of any system or plant, its behaviour—as represented by its transfer function $G(s)$—may change as a result of changes in the plant parameters, such as flow rates, pressures and temperatures, heat-transfer coefficients, thermal conductivities and the effects of ageing in general. It is essential to ensure, as far as possible, that the effect of any change in the nominal transfer function has a minimum effect upon the behaviour of the closed-loop system and it is important that this so-called *sensitivity* is considered at the design stage.

Steady-state and Transient Behaviour of Control Systems

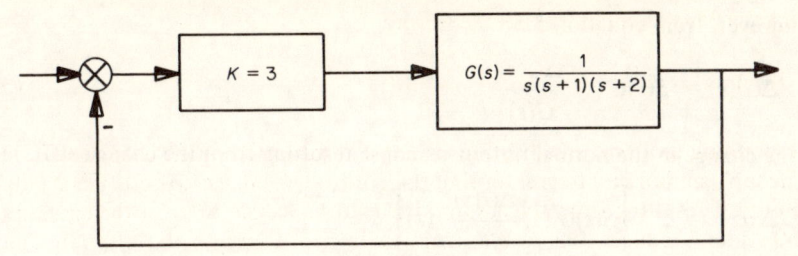

Figure 5.21

To illustrate this effect, consider the control system of figure 5.21, which has been successfully designed and is deemed to operate satisfactorily at the outset. Suppose now that, following some time in operation, the transfer function changes to

$$\frac{1}{s(s+1)(s+1.2)}$$

It is a simple matter to show, using the Routh method, that this modified system will be unstable. Thus, it is essential to predict any future changes that may take place, either due to ageing or change in operating policies, and to design the control system using the worst possible case.

Consider the case where only small changes are permitted in the system transfer function. If $\Delta G'(s)$ is the change in $G'(s)$ resulting from a small change $\Delta G(s)$ in $G(s)$, then it follows from the relation

$$G'(s) = \frac{KG(s)}{1 + KG(s)} \qquad (5.55)$$

Tayler Expansion

that

$$\Delta G'(s) = \frac{K\Delta G(s)}{1 + KG(s)} - KG(s) \frac{K\Delta G(s)}{[1 + KG(s)]^2} \qquad (5.56)$$

neglecting second-order effects. Therefore

$$\Delta G'(s) = \frac{\Delta G(s)}{G(s)} G'(s) - G'(s) \frac{\Delta G(s)}{G(s)} G'(s) \qquad (5.57)$$

and

$$\frac{\Delta G'(s)}{G'(s)} = [1 - G'(s)] \frac{\Delta G(s)}{G(s)} = \frac{1}{1 + KG(s)} \frac{\Delta G(s)}{G(s)} \qquad (5.58)$$

Thus, the sensitivity of the closed-loop control system S, defined by

$$S(s) = \frac{\Delta G'(s)/G'(s)}{\Delta G(s)/G(s)} \qquad (5.59)$$

is equal to $1/[1 + KG(s)]$. It is difficult to interpret this result in the time domain but possible in the frequency domain. More will be said about this later.

However, from equation 5.58

$$\Delta G'(s) = S(s)G'(s) \frac{\Delta G(s)}{G(s)} \tag{5.60}$$

and the change in the normal output response resulting from the change $\Delta G(s)$ is

$$\Delta x(t) = \mathcal{L}^{-1}\left[S(s)G'(s)\frac{\Delta G(s)}{G(s)} r(s)\right] \tag{5.61}$$

where $r(s)$ is the Laplace transform of the reference input. Equation 5.61 appears complicated but in many cases judicious cancelling of factors may lead to a simple expression for $\Delta x(t)$ and hence for the effect of the change.

Further, equation 5.56 may be written in the form

$$\Delta G'(s) = \frac{K}{1 + KG(s)} \frac{1}{1 + KG(s)} \Delta G(s) \tag{5.62}$$

and $\Delta x(t)$ may be obtained from the system shown in figure 5.22.

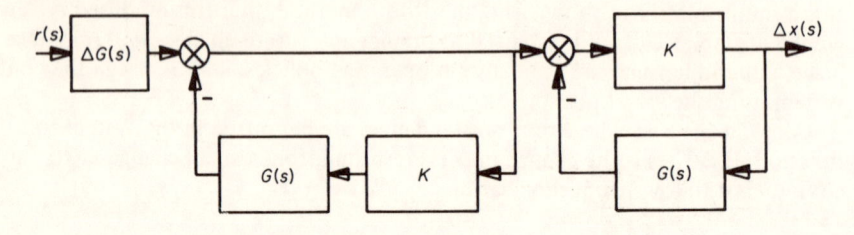

Figure 5.22

5.6 BEHAVIOUR OF A GOVERNOR SYSTEM

To illustrate the principles outlined above, consider the following example.

Example 5.10
The spring-loaded governor shown in figure 5.23 is operating at rated speed. Assume that friction in the linkages, sleeve inertia, viscous friction, etc., can be neglected.

(a) Show that, for small deviations, the output motion of the governor Δh is approximately proportional to the variation in speed, and evaluate this ratio in terms of the spring constant K_s at steady rated conditions, the rated turbine speed being 60π rad/s.

(b) The steam-valve setting α, is related linearly to h in such a way that $\alpha = 90°$ when $h = 0$ and $\alpha = 45°$ when $h = h_0$. The internal developed turbine torque M is a function of α only, such that, at rated conditions, a $5°$ change in α causes a change of 10 per cent in M. The turbine-generator set has a moment of inertia I of 4.071 kg m^2 and its rotation is opposed by viscous damping of μ times its angular velocity, where $\mu = 1.356$ N m/rad/s. The load torque M_L is proportiona to the generator output: the generator efficiency is 95 per cent and its rated

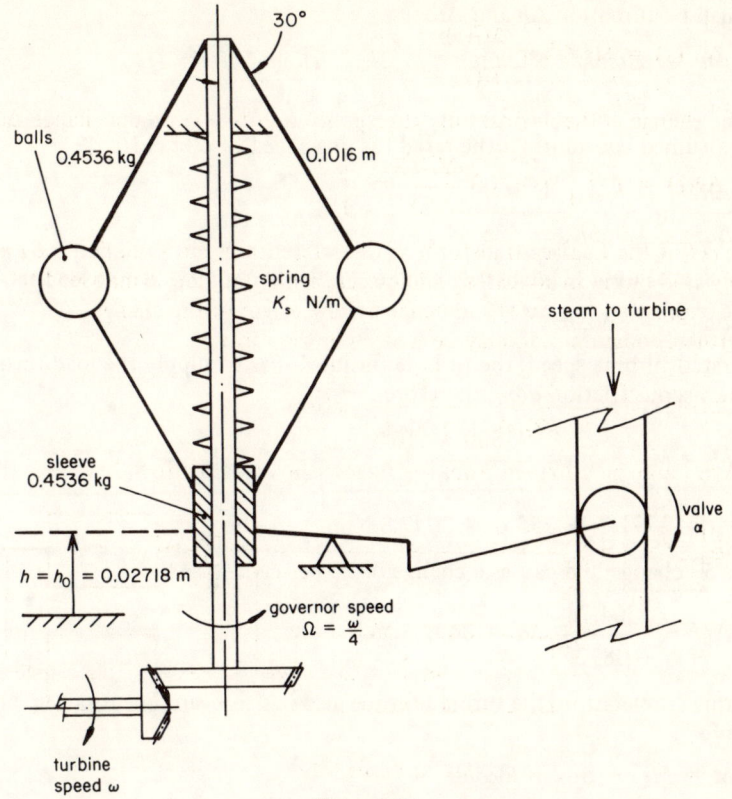

Steady-state and Transient Behaviour of Control Systems

![Figure 5.23 - Governor diagram with balls (0.4536 kg), 30° angle, 0.1016 m, spring K_s N/m, sleeve 0.4536 kg, $h = h_0 = 0.02718$ m, governor speed $\Omega = \omega/4$, steam to turbine, valve a, turbine speed ω]

Figure 5.23

output is 500 kW. Find the turbine torque under rated conditions. Determine the value of K_s such that, following a small step disturbance in M_L, at rated speed, the speed reaches 98 per cent of its new equilibrium value in 1 s. Also find equations for turbine speed and torque following a sudden overload of 10 per cent on the generator.

(c) Using the value of K_s already calculated plot $\Delta h/\Delta \omega$ against ω for speeds between 43.33π and 76.67π rad/s. Is it correct to assume, as in (a), that $\Delta h/\Delta \omega$ is practically constant over the speed range?

(a) Considering a small displacement δh of the sleeve, the principle of virtual work leads to the relation

$$(l - \tfrac{1}{2}h)m_b\Omega^2 = m_s g + m_b g + (F_0 + K_s h)g$$

Therefore

$$h = \frac{l\Omega^2 m_b - (m_s g + m_b g + F_0)}{K_s + \tfrac{1}{2}m_b \Omega^2}$$

$$= \frac{l\omega^2 m_b - 16(m_s g + m_b g + F_0)}{16 K_s + \tfrac{1}{2}m_b \omega^2}$$

Introduction to Control Theory

For small perturbation Δh and $\Delta \omega$

$$\frac{\Delta h}{\Delta \omega} = \frac{2l\omega m_b - \omega h m_b}{16K_s + \frac{1}{2}m_b \omega^2} = \frac{(2l - h)\omega m_b}{16K_s + \frac{1}{2}\omega^2 m_b}$$

and is assumed constant. At the rated turbine speed of 60π rad/s

$$\frac{\Delta h}{\Delta \omega} = \frac{1}{1.063 K_s + 535.44}$$

(b) Given that $\alpha = \pi/2$ when $h = 0$ and $\alpha = \pi/4$ when $h = h_0$, it follows that

$$\Delta \alpha = -28.9 \Delta h$$

At rated turbine speed, the turbine torque M has to supply the load torque M_L and the viscous friction $\mu\omega_0$; therefore

$$M = M_L + \mu\omega_0 = \frac{500 \times 10^3}{0.95} \times \frac{1}{188.4} + 188.4 \times 1.356$$

$$= 2792.2 + 255.6 = 3047.8 \text{ N m}$$

Since a $5°$ change in α causes a change of 10 per cent in M

$$\Delta M = \frac{304.78}{0.08727} \Delta \alpha = 3492.5 \Delta \alpha$$

During acceleration, the turbine torque also has to overcome rotational inertia; therefore

$$M = M_L + \mu\omega + I d\omega/dt$$

and assuming small changes

$$\Delta M = \Delta M_L + \mu \Delta \omega + I \frac{d\Delta \omega}{dt}$$

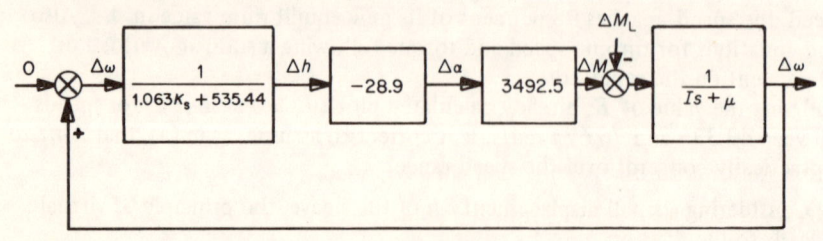

Figure 5.24

Figure 5.24 shows a block diagram of the system: note the sign convention.

Therefore, assuming zero initial conditions (this is correct since the equations are written in terms of perturbed variables)

$$-\left[3492.5 \times 28.9 \times \frac{1}{1.063 K_s + 535.44}\right]\Delta\omega(s) = \Delta M_L(s) + 1.356 \Delta\omega(s)$$

$$+ 4.071 s \Delta\omega(s)$$

Steady-state and Transient Behaviour of Control Systems

Therefore

$$\Delta\omega(s) = -\frac{\Delta M_L(s)}{4.071s + 1.356 + A}$$

Now ΔM_L is a small step disturbance; therefore

$$\Delta\omega(s) = \frac{-\Delta M_L}{s(4.071s + 1.356 + A)} \qquad \frac{\Delta M_L/(1.356 + A)}{s} + \frac{\Delta M_L/(1.356 + A)}{s + (1.356 + A)/4.071}$$

and

$$\Delta\omega(t) = -\frac{\Delta M_L}{1.356 + A}\left(1 - e^{-[(1.356 + A)/4.071]t}\right)$$

The speed reaches 98 per cent of its new equilibrium value after 1 s; therefore

$$1 - e^{-(1.356 + A)/4.071} = 0.98$$

Hence

$$A = 14.57$$

and

$$K_s = 6013.27 \text{ N/m}$$

For an overload of 10 per cent, $\Delta M_L = 279.22$ N m. Therefore

$$\Delta\omega(t) = \frac{279.22}{15.926}(1 - e^{-3.91t})$$

Therefore the turbine speed following overload is

$$60\pi - 17.53(1 - e^{-3.91t}) \text{ rad/s}$$

The change in turbine torque is

$$\Delta M = \Delta M_L + \mu\Delta\omega + I\frac{d}{dt}\Delta\omega$$

$$= 255.45 - 255.27 e^{-3.91t}$$

Therefore the total turbine torque following overload is

$$3047.8 + 255.45 - 255.27 e^{-3.91t} \text{ N m}$$

(c) A plot of $\Delta h/\Delta\omega$ against ω is nearly linear over the range from 43.33π to 76.67 rad/s, and hence the assumption is justified. The reader should verify this.

5.7 USE OF COMPUTER GRAPHICS

It will be evident that to obtain the transient response of a closed-loop control system for various values of gain K, taking into account the effect of velocity feedback, derivative action and integral action, is extremely laborious and obviously great care must be taken to avoid errors and such like.

This procedure can be automated and hence a great many more values and modifications can be taken into account by the use of a digital computer that has an interface to a graphics terminal. The initial input to the computer is the transfer function, which is followed by information concerning the control configurations and the ranges of parameters to be considered. The program then evaluates the closed-loop transfer function and, assuming a step input, obtains the time response by either an inverse-Laplace-transform subroutine or matrix methods. This time response then forms the output to the graphics terminal.

Figure 5.25 shows the results of using such a computer package to design a suitable controller for the open-loop system

$$G(s) = \frac{6}{(s+1)(s+2)(s+3)}$$

REFERENCES

Murphy, G. J., *Basic Automatic Control Theory* (Van Nostrand, New York, 1966).
Routh, E. J., *Stability of Motion*, ed. A. T. Fuller (Taylor and Francis, London, 1975).
Savant, C. J., *Basic Feedback Control System Design* (McGraw-Hill, New York, 1958).

PROBLEMS

5.1 Discuss the performance requirements of a gun-control system, an air-conditioning system, an oil-refinery and power-assisted steering.

5.2 Use Routh to determine the range of values of the gain K for which a unity-feedback system with forward transference

$$KG(s) = \frac{K(s^2 + s + 1)}{s(s^2 + s + 4)(s+2)(s+5)}$$

is stable. Calculate the two equal and opposite roots for a particular value of the gain.

5.3 The denominator of a closed-loop transfer function is

$$s^6 + 2s^5 + 4s^4 + 8s^3 + s^2 + 10s + 1$$

Determine whether the system is stable or unstable; if it is unstable, determine the number of roots with positive real parts.

5.4 Use Routh to determine the stability of a system with characteristic equation

$$s^5 + 2s^4 + 6s^3 + 12s^2 + s + 2 = 0$$

5.5 Determine the position, velocity and acceleration error constants for the unity-feedback system with forward transference

$$\frac{2(s+2)}{s^2(s+4)}$$

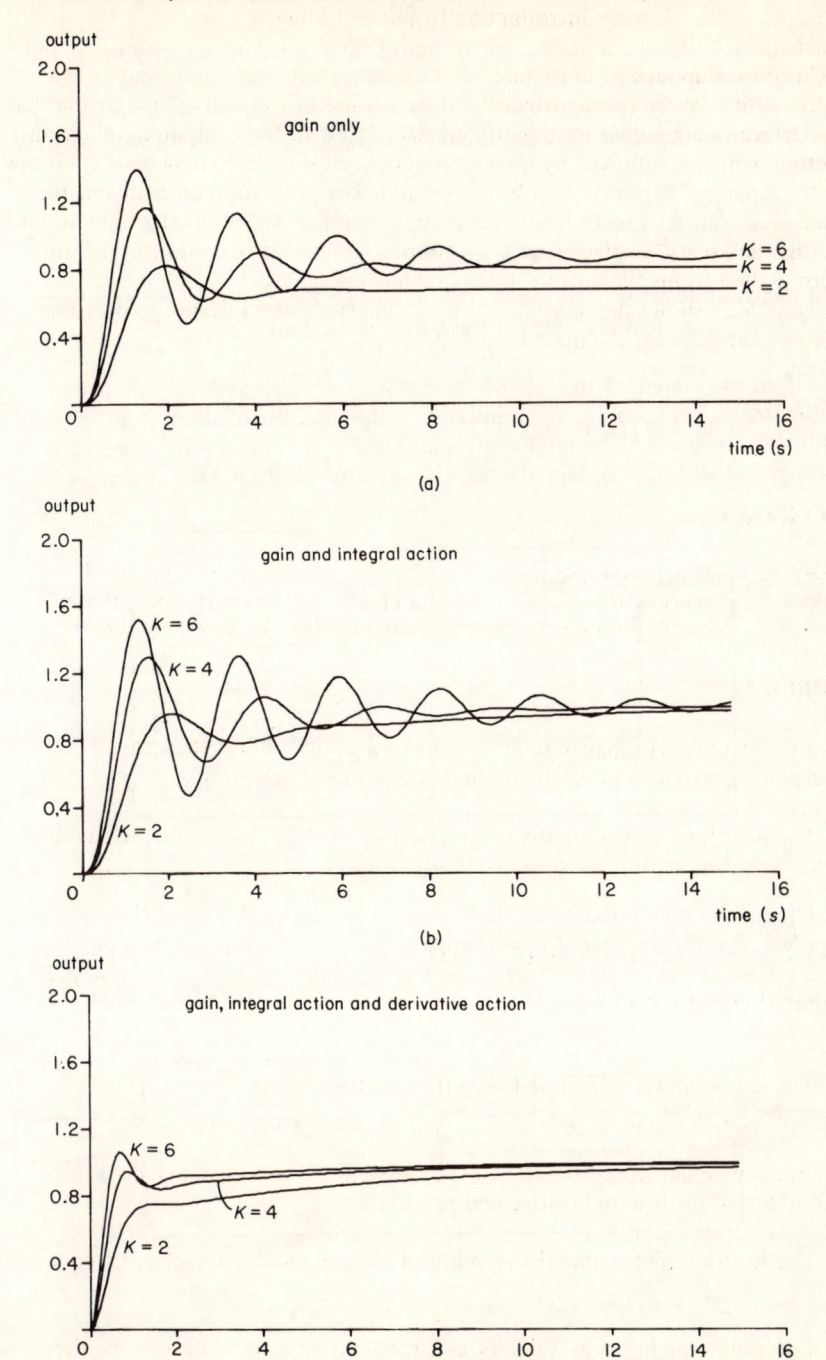

Steady-state and Transient Behaviour of Control Systems

Figure 5.25

Hence deduce the steady-state error for unit step input, unit velocity input and unit acceleration input.

5.6 Determine the value of K such that the unity-feedback system with forward transference

$$\frac{K}{s^2 + 3s + 2}$$

has a steady-state error of not more than 0.2 when excited by a unit step input. Check that the system is stable for this value of the gain.

5.7 Given the system of figure 5.26, determine the value of T such that the predominant time constant is 5 s and obtain the response of the system following a unit step change in the reference signal when $K = 0.3$ and $\alpha = 0.1$. Compare this response with that obtained when the velocity-feedback loop is removed.

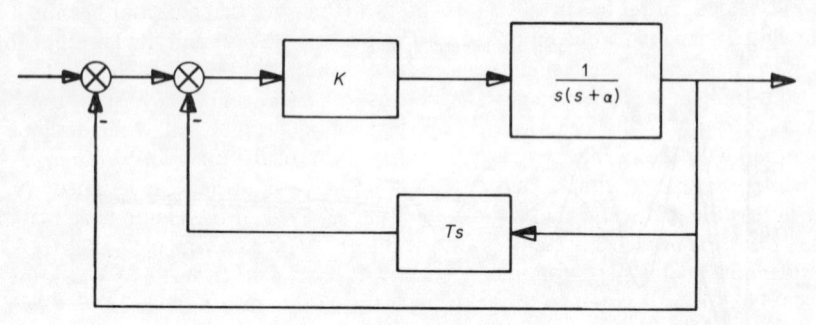

Figure 5.26

5.8 Obtain the time response of the system shown in figure 5.27 after a unit step is applied, for integral time $T = \infty$. Compare this with the response obtained when $T = 2$ s. Discuss the system behaviour as T is gradually reduced to zero. Is $T = 2$ s a suitable choice for the integral time?

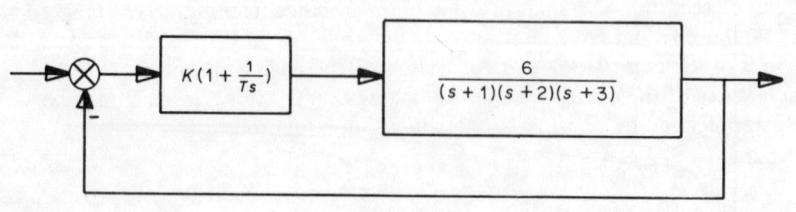

Figure 5.27

6 The Root-locus Method

So far, the meaning, use and representation of a closed-loop control system have been discussed, and it has been established how to derive its behaviour in time following a change in some input signal. The present chapter and the two that follow go on to consider the analysis and design (or synthesis) of these systems. In this context, analysis is concerned with the means by which, given the form of the closed-loop system, the behaviour of the system may be analysed, while design is concerned with the means by which, given the form of the open-loop system, a controller that produces satisfactory closed-loop behaviour may be designed.

Whereas the preceding chapter examined the analysis of the closed-loop system in the *time domain* (this is equivalent to setting $s = d/dt$) and the next chapter considers the analysis in the *frequency domain* (equivalent to setting $s = j\omega$), the present chapter is devoted to the analysis of the system over the whole *complex plane* (equivalent to setting $s = \sigma + j\omega$). In chapter 8 these techniques of analysis are used in the design of a closed-loop system.

Although these techniques are presented as if they were self-contained, in practice it is usual to apply either the complex-plane or frequency-domain technique in the initial stages of the analysis or design of a system and then to check its behaviour in the time domain. In fact, this procedure is recommended and is adopted in the examples presented throughout the remainder of the book.

It will be seen that the analysis is based on graphical techniques (suitable for hand calculation) that are able to provide information on the stability and behaviour of the closed-loop control system on the basis of knowledge of the characteristics of the open-loop system. The reason for this approach may be readily appreciated by considering the closed-loop relationship

$$G'(s) = \frac{KG(s)}{1 + KG(s)} \tag{6.1}$$

Usually $G(s)$, the open-loop transfer function, is in factored form and $G'(s)$, because of the form of equation 6.1, is not. As was seen earlier, it is relatively simple to obtain a transient response using partial fractions when the transfer function is factored but conversely not quite so straightforward when it is not factored. Hence there is a need for techniques from which the closed-loop system behaviour can be obtained.

Carrying on this argument a little further, it may be claimed that the availability of digital computers with 'hands-on' display facilities has made these hand graphical techniques obsolete, but this is not so. Such techniques form an important part of the understanding of the design procedures, which is essential, and in fact the two approaches fully complement one another. Examples of the use of digital computers are given at the end of the chapter.

Evans (1948) has developed a method of analysing the behaviour of a closed-loop control system called the root-locus method; this is a graphical technique by means of which the locus (or variation) of the roots of the characteristic equation of the closed-loop system produced by the variation of some system parameter—usually the system gain K—can be deduced from knowledge of the open-loop system together with a set of rules for constructing the locus. From this locus it is possible to choose a particular value of K that is likely to result in reasonable stability, and it is then quite simple to obtain the closed-loop transient response. In fact, this is a graphical method of factorising the closed-loop transfer function $G'(s)$.

Before proceeding to the discussion of the root-locus method, it is necessary to discuss the meaning and effect of the poles and zeros of the open-loop transfer function $G(s)$.

6.1 POLE-ZERO CONFIGURATION

The open-loop transfer function $G(s)$ is written as a ratio of two polynomials in s

$$G(s) = \frac{b_0 s^m + b_1 s^{m-1} + \ldots + b_m}{a_0 s^n + a_1 s^{n-1} + \ldots + a_n} \tag{6.2}$$

in which the order of the numerator is less than the order of the denominator, that is, $m < n$, and can be factorised to give

$$G(s) = K' \frac{(s - z_1)(s - z_2) \ldots (s - z_m)}{(s - p_1)(s - p_2) \ldots (s - p_n)} \tag{6.3}$$

where $K' = b_0/a_0$. The roots of the numerator, $z_1, z_2, \ldots, z_m$, are called the zeros of $G(s)$ since setting s equal to any of these values makes $G(s)$ zero. The total number of zeros is m (counting a zero which repeats itself l times as l zeros). The roots of the denominator, $p_1, p_2, \ldots, p_n$, are called the poles of $G(s)$ since setting s equal to any of these values makes $G(s)$ infinite. There are n poles and they are identical to the roots of the characteristic equation of the system with transfer function $G(s)$. The poles and zeros of $G(s)$ can be real and complex, but if complex they must occur in conjugate pairs, that is, $p_1 = \sigma_1 + j\omega_1$ and $p_2 = \sigma_2 - j\omega_2$, since the polynomials of equation 6.2 are real. Also, if the transfer function represents a stable system, the real parts of each of the n poles must be negative (cf. the real parts of the roots of the characteristic equation).

Knowledge of the n poles and m zeros of $G(s)$ allows the latter to be conveniently represented on the complex plane. The poles and zeros are represented as crosses and circles respectively and the pattern so formed is known as the *pole-zero configuration* of $G(s)$. For example, consider the system represented by

The Root-locus Method

$$G(s) = \frac{s^2 + s}{s^3 + 8s^2 + 21s + 20} \tag{6.4}$$

which may be factorised to give

$$G(s) = \frac{s(s + 1)}{(s + 4)(s + 2 + j1)(s + 2 - j1)} \tag{6.5}$$

This system is seen to have two zeros, $z_1 = 0$ and $z_2 = -1$, and three poles, $p_1 = -4$, $p_2 = -2 + j1$ and $p_3 = -2 - j1$; the pole–zero configuration is shown in figure 6.1.

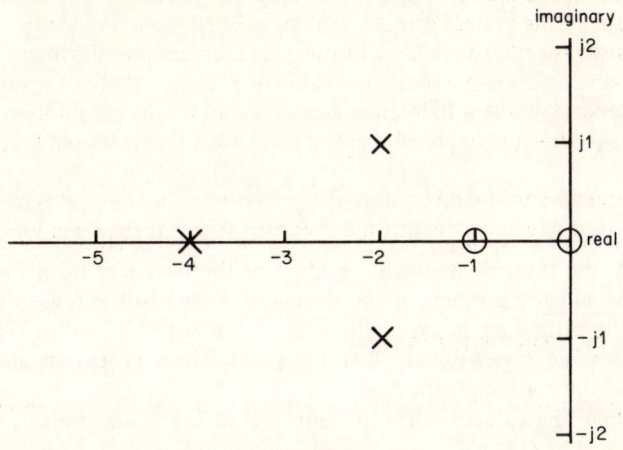

Figure 6.1

Note that since the complex poles (and zeros) appear as conjugate pairs the configuration is symmetrical about the real axis. Also, since the three poles have negative real parts, they all appear in the left half-plane of the complex plane and the system is therefore stable.

Each of the factors of equation 6.3 is in general a complex quantity, or may be represented as one, and hence may be expressed in an alternative form (known as *polar form*). Let

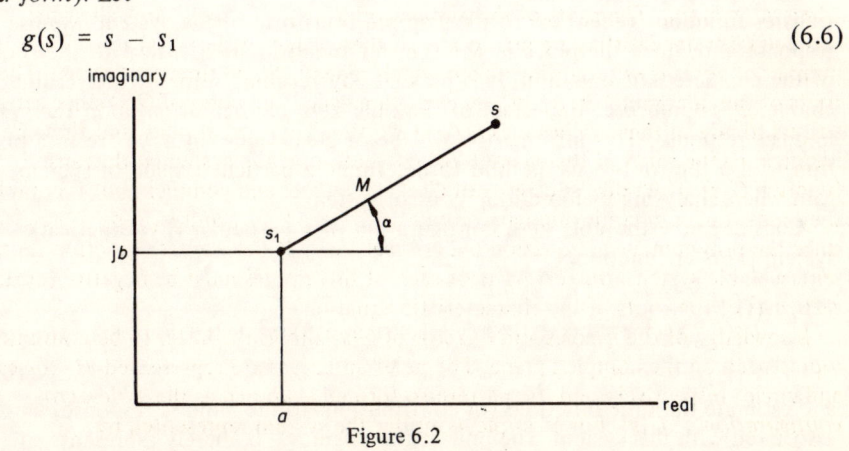

$$g(s) = s - s_1 \tag{6.6}$$

Figure 6.2

as shown in figure 6.2; then

$$g(s) = s - a - jb = Me^{j\alpha} \tag{6.7}$$

where $M = \text{mod}\,[g(s)] = \{[R(s) - a]^2 + [I(s) - b]^2\}^{1/2}$, that is, the distance between s and s_1, and $\alpha = \arg\,[g(s)] = \tan^{-1}\,\{[I(s) - b]/[R(s) - a]\}$, that is, the angle between the line ss_1 and the real axis, positive being anti-clockwise. Thus in general

$$G(s) = K' \left[\frac{M_1 e^{j\alpha_1}\, M_2{}^{j\alpha_2}\,\ldots\, M_m e^{j\alpha_m}}{N_1 e^{j\beta_1}\, N_2 e^{j\beta_2}\,\ldots\, N_n e^{j\beta_n}}\right] \tag{6.8}$$

$$= Me^{j\gamma}$$

$$= K'\,|G(s)|\arg\,[G(s)]$$

$$= K'\,|G(s)|\angle G(s) \tag{6.9}$$

$$= K'\,|G(s)|\angle\gamma(s)$$

where $M = K'M_1 M_2 \ldots M_m/N_1 N_2 \ldots N_n$ and is simply the constant K' multiplied by the ratio between the product of the distances from the given point s to the zeros and the product of the distances from s to the poles. The angle γ is simply the difference between the sum of the angles extended to the zeros and the sum of the angles extended to the poles. These properties are used extensively in the root-locus method.

For example, the value of $G(s)$ of figure 6.1 at $s = -3$ is given by

$$G(s) = 1 \times \frac{3e^{(j\pi)} \times 2e^{(j\pi)}}{1e^{(j0)} \times \sqrt{2}e^{(j\pi + \pi/4)} \times \sqrt{2}e^{(j\pi - \pi/4)}} \tag{6.10}$$

$$= 1 \times \frac{3 \times 2}{1 \times \sqrt{2} \times \sqrt{2}} \times e^{(j\pi + j\pi - j0 - \overline{j\pi + \pi/4} - \overline{j\pi - \pi/4})}$$

$$= 3$$

Since the Laplace transform of an impulse response is unity, the system transfer function is identical to the Laplace transform of the system impulse response and, since the poles of the transfer function are identical to the roots of the characteristic equation, it is possible, by recalling some of the results of chapter 3, to interpret the effect of the pole–zero configuration upon the system impulse response. This information will be of great value once we are in a position to use the root-locus method to determine a particular value of system gain when analysing a closed-loop control system.

Consider now the pole–zero configuration of a particular (hypothetical) system transfer function

$$G(s) = \frac{s + 2}{s(s^2 + 1)(s + 0.5)(s + 1.4)(s + 1.6)(s^2 + 2s + 5)(s + 2.1)(s + 3)} \tag{6.11}$$

and estimate the time functions (or contributions to the impulse response) associated with this system. The pole at the origin, $s = 0$, merely represents an integration in time. The pole at $s = -0.5$ gives rise to a time function with a slow exponential decay and that at $s = -3.0$ to a function with a much faster decay. The two poles close together on the real axis at $s = -1.4$ and $s = -1.6$ can be approximated by a pair of poles at $s = -1.5$; their time functions will usually be of opposite sign. The pole at $s = -2.1$ contributes very little to the time function because of the nearness of the zero at $s = -2.0$ and so pole–zero pairs may be neglected. The conjugate pair of poles on the imaginary axis at $s = \pm j1$ gives rise to an oscillatory time function, whose frequency will increase as the poles move further along the imaginary axis away from the origin. The conjugate pair at $s = -1 \pm 2j$ gives rise to a damped oscillatory time function. In general a conjugate pair of poles at $s = -a \pm jb$ gives rise to a damping factor $a/\sqrt{(a^2 + b^2)}$ and an undamped natural frequency $\sqrt{(a^2 + b^2)}$.

6.2 ROOT-LOCUS METHOD

The root-locus method, which is based upon a straightforward graphical method of determining the closed-loop-system poles from knowledge of the open-loop-system poles and zeros and hence determining the closed-loop-system transient response, gives considerable insight into the behaviour of the closed-loop system. Thus given the system of figure 6.3 the root-locus method can be used to determine a value for the controller K such that the system exhibits reasonable transient behaviour.

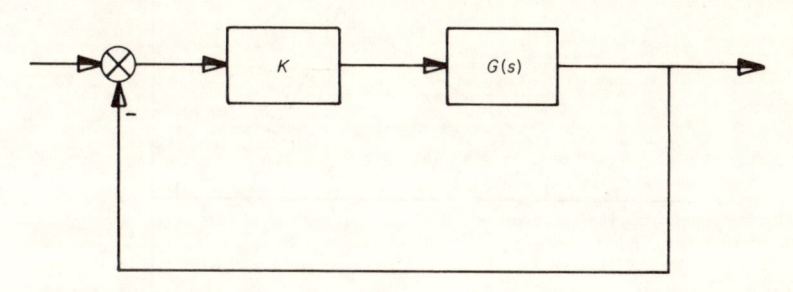

Figure 6.3

Consider now the closed-loop transfer function

$$G'(s) = \frac{KG(s)}{1 + KG(s)} \tag{6.12}$$

in which the closed-loop poles occur at the values of s for which the denominator of $G'(s)$ is zero

$$1 + KG(s) = 0 \tag{6.13}$$

or

$$KG(s) = -1 \tag{6.14}$$

From equation 6.7 this can be expressed in polar form as

$$KG(s) = 1e^{j\pi} \tag{6.15}$$

or, more completely, as

$$KG(s) = 1e^{j(2n+1)\pi} \quad n = 0, 1, 2, \ldots \tag{6.16}$$

This can be written as two separate equations

$$\left. \begin{array}{l} \text{mod } [KG(s)] = 1 \\ \text{arg } [KG(s)] = (2n+1)\pi \quad n = 0, 1, 2, \ldots \end{array} \right\} \tag{6.17}$$

that is, an odd multiple of $180°$.

Thus the diagram constructed in the root-locus method is a plot of all the values of s that satisfy the angle criterion of equation 6.17 and hence it represents the loci of the closed-loop poles, that is, the roots of equation 6.14—for all values of K as K varies from zero to infinity. Using this diagram it is then possible by satisfying the magnitude criterion of equation 6.17, to select a particular value of K that gives rise to a closed-loop system with a known set of poles.

The construction of the locus of points that satisfy the angle criterion is accomplished directly on the complex s plane. The poles and zeros of $KG(s)$, that is, the open-loop system, are initially placed on the complex plane, and the root-locus is obtained by determining graphically all values of s that satisfy the angle

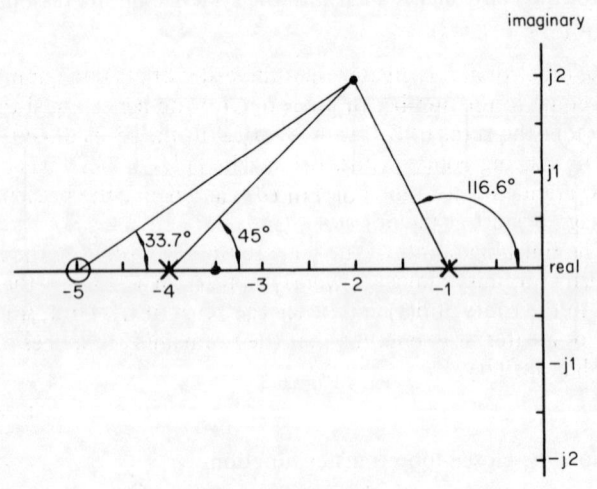

Figure 6.4

criterion. Consider for example the pole–zero configuration of figure 6.4. At the point $s = -2 + j2$, arg $[KG(s)]$ is given by

$$\text{arg } [KG(s)] = 33.7 - 116.6 - 45 = -127.9°$$

which does not satisfy the angle criterion and therefore does not lie on the root locus; hence $s = -2 + j2$ is not a pole of the corresponding closed-loop system. Consider now the point $s = -3.5$, where $\text{arg}[KG(s)] = 0 - 180 - 0 = -180°$, which is an odd multiple of $180°$; $s = -3.5$ lies on the root-locus and is a pole of the closed-loop system corresponding to a particular value of gain K. Using the magnitude criterion, this value is given by

$$K = \frac{1}{|G(s)|} = 1 \bigg/ \frac{\text{(product of distances to zeros)}}{\text{(product of distances to poles)}}$$

$$= \frac{2.5 \times 0.5}{1.5} = 0.83$$

Obviously proceeding in point-by-point, trial-and-error fashion is very time-consuming and there is evidently a need for a method for the quick determination of these loci.

6.3 RULES FOR DRAWING THE ROOT-LOCUS DIAGRAM

The following rules are useful in determining the loci of the root-locus diagram.
Rule 1: The loci are symmetrical about the real axis.
Rule 2: The number of loci is equal to the number of poles of $G(s)$. Let

$$G(s) = b(s)/a(s)$$

Then

$$G'(s) = \frac{Kb(s)}{a(s) + Kb(s)} \qquad (6.18)$$

and, since the order of $a(s)$ is greater than the order of $b(s)$, the number of poles of $G'(s)$ will be equal to the number of poles of $G(s)$ and hence equal to the number of loci. Note that the zeros of $G'(s)$ are identical to the zeros of $G(s)$ at all times.

Rule 3: The starting points of the loci are the poles of $G(s)$. When $K = 0$, the poles of $G'(s)$ are identical to the poles of $G(s)$ and hence the branches of the root-locus diagram start at the poles of $G(s)$.

Rule 4: The finishing points of the loci are the zeros of $G(s)$; the remaining branches finish at infinity along known asymptotes. When $K = \infty$ the poles of $G'(s)$ are identical to the roots of $b(s)$ and hence the zeros of $G(s)$ but, since the rank r of $G(s)$ is unity or greater, it is assumed that the r remaining branches of the root-locus diagram finish at infinity.

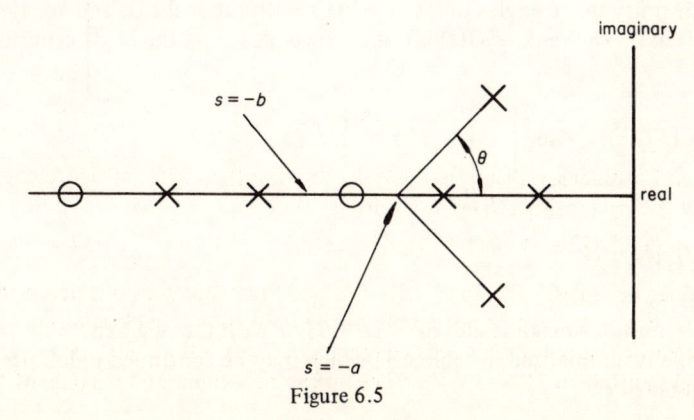

Figure 6.5

Rule 5: A point on the real axis lies on the root locus if there is an odd number of poles and zeros to the right of it. Consider the pole–zero configuration of figure 6.5 and use the angle criterion to determine whether firstly $s = -a$ and secondly $s = -b$ lie on the loci.

In this case

$$\arg[KG(s)]_{s=-a} \quad 0 + 0 - \pi - \pi - 0 - 0 - (\pi + \theta) - (\pi - \theta)$$
$$= -4\pi$$

and therefore $s = -a$ does not lie on the loci. Note that the last two terms are due to the complex pair of poles and since their contribution is -2π they can be neglected in this instance. Now

$$\arg[KG(s)]_{s=-b} = \pi + 0 - \pi - \pi - 0 - 0 = -\pi$$

and hence $s = -b$ does lie on the loci; the loci are shown in figure 6.6. The reader should confirm that points $s = -c$ and $s = -e$ do lie on the loci and $s = -d$ does not.

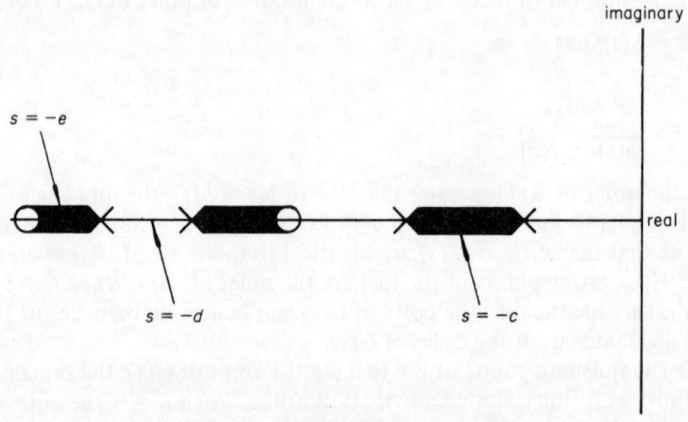

Figure 6.6

Rule 6: The asymptotes along which the r remaining branches of the root locus finish at infinity make angles of $-(2n + 1)\pi/r$ with the real axis and are symmetrical about this axis. The rank of $G(s)$ is r and hence, as $s \to \infty$, the angle criterion becomes

$$\arg[KG(s)] = \arg\left[\frac{1}{s^r}\right] = (2n + 1)\pi \qquad (6.19)$$

Therefore

$$\arg(s) = \frac{-(2n + 1)\pi}{r} \qquad (6.20)$$

and the asymptotes make angles of $-(2n + 1)\pi/r$ with the real axis.

This rule is summarised in table 6.1, which may be continued indefinitely for higher values of r.

The Root-locus Method

Table 6.1

Value of r	Number of Asymptotes	Inclination to Real Axis
1	1	$-180°$
2	2	$-90°, 90°$
3	3	$-60°, -180°, 60°$
4	4	$-45°, -135°, 45°, 135°$

Rule 7. The point of intersection of the asymptotes on the real axis is given by ($\Sigma poles - \Sigma zeros$)/rank. Let

$$KG(s) = K \frac{s^m + b_1 s^{m-1} + \ldots + b_m}{s^n + a_1 s^{n-1} + \ldots + a_n} \quad (6.21)$$

$$= \frac{K}{s^r + (a_1 - b_1)s^{r-1} + \ldots} \quad (6.22)$$

where $r = n - m$, and therefore using equation 6.17 again

$$s^r + (a_1 - b_1)s^{r-1} + \ldots = -K \quad (6.23)$$

Since, for the case when the root loci and the asymptotes meet, s must be large, equation 6.23 can be approximated by

$$\left(s + \frac{a_1 - b_1}{r}\right)^r = -K \quad (6.24)$$

Therefore, at the point of intersection of the asymptotes $K = 0$ and

$$s = \frac{b_1 - a_1}{r} = \frac{\Sigma poles - \Sigma zeros}{r} \quad (6.25)$$

(Remember that for the equation $c_n x^n + c_{n-1} x^{n-1} + c_{n-2} x^{n-2} + \ldots + c_0 = 0$, the sum of the roots is $-c_{n-1}/c_n$.)

For example, consider the open-loop system with the pole–zero configuration shown in figure 6.7. The rank is four and so the asymptotes are at angles $-45°$, $-135°$, $45°$, $135°$ and intersect at

$$s = \frac{[(-1) + (-3) + (-6) + (-4.5 - j2) + (-4.5 + j2)] - [-8]}{4}$$

$$= \frac{-11}{4} = -2.75$$

The asymptotes are shown dashed in figure 6.7.

Rule 8. This is concerned with departure from and arrival to the real axis. When there is a pole at each end of a locus on the real axis, the two poles move towards each other as K increases, forming a double pole, and then break away (initially at right angles) to form a pair of conjugate poles. The converse situation occurs

114   Introduction to Control Theory

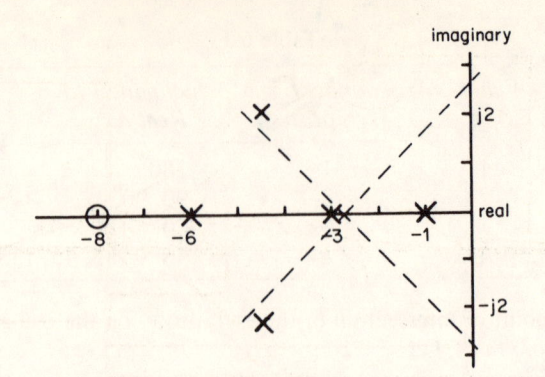

Figure 6.7

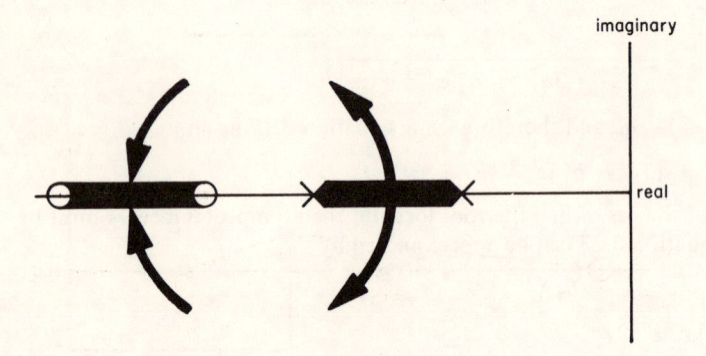

Figure 6.8

when there is a zero at each end of a locus, as shown in figure 6.8; similar calculations may be carried out to find points of departure and arrival.

Consider the pole-zero configuration shown in figure 6.9, where it is required to calculate the point of departure of the locus between the two right-hand poles on the real axis: let the point be at $-z$. Take a point very near the real axis at $-z + j\delta$; then from the angle criterion (remembering the sign convention for angles)

$$\theta_1 - \theta_2 - \theta_3 - (180 - \theta_4) - (\theta_5 + \alpha_5) - (\theta_6 + \alpha_6) = 180°$$

that is

$$\theta_1 - \theta_2 - \theta_3 + \theta_4 - \alpha_5 - \alpha_6 = 0 \qquad (6.26)$$

since

$$\theta_5 + \theta_6 = 360°$$

Now since δ is assumed small, the angles can be replaced by their tangents, to give

$$\frac{\delta}{x_5} - \frac{\delta}{x_4} - \frac{\delta}{x_2} + \frac{\delta}{x_1} - \frac{2\delta x_3}{x_3^2 + y_3^2} = 0 \qquad (6.27)$$

from which the position of the breakaway z may be calculated.

The Root-locus Method

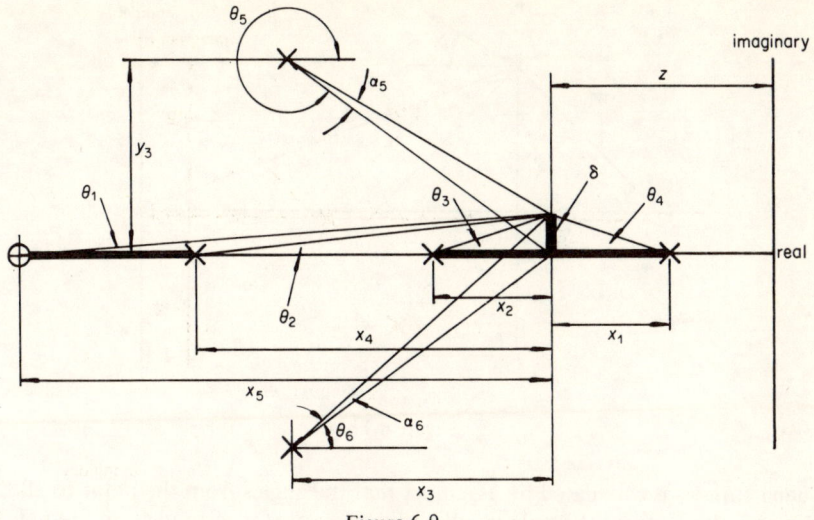

Figure 6.9

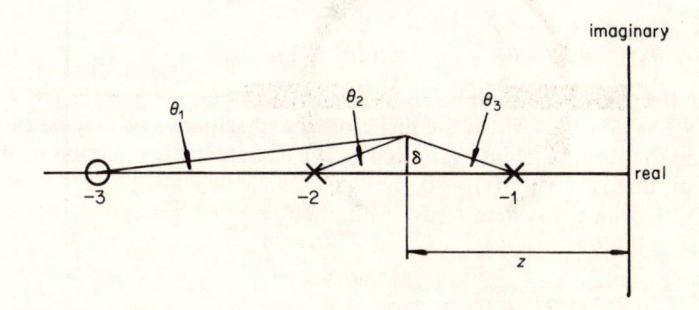

Figure 6.10

For the system of figure 6.10 this becomes

$$\theta_1 - \theta_2 + \theta_3 = 0$$

Therefore

$$\frac{\delta}{3-z} - \frac{\delta}{2-z} + \frac{\delta}{z-1} = 0$$

and

$$z^2 - 6z + 7 = 0$$

Hence

$$z = 1.586$$

Rule 9 This covers departure from complex poles. The angle of departure from a complex pole (or arrival at a complex zero) is obtained by application of the angle criterion. Consider a point very near to the complex pole p in figure 6.11. The angle

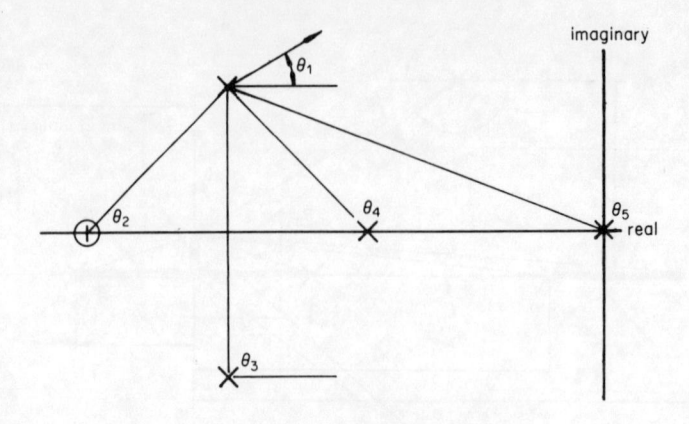

Figure 6.11

of departure θ_1 is calculated by assuming that the angles from the point to all the poles and zeros except the pole p will be the same as if measured from the pole p; thus

$$\theta_2 - \theta_3 - \theta_4 - \theta_5 - \theta_1 = (2n + 1)\pi \qquad (6.28)$$

and hence the angle θ_1 is found.

Rule 10 The points at which the loci cross the imaginary axis may be calculated by applying the Routh stability criterion to the characteristic equation of the closed loop system, that is, $1 + KG(s) = 0$.

For example, let the system transfer function be

$$G(s) = \frac{1}{(s + 3)(s^2 + 2s + 2)}$$

Then

$$1 + \frac{K}{(s + 3)(s^2 + 2s + 2)} = 0$$

and

$$(s + 3)(s^2 + 2s + 2) + K = 0$$

or

$$s^3 + 5s^2 + 8s + 6 + K = 0$$

which is the characteristic equation of the closed-loop system.

Applying the Routh criterion gives the array

$$
\begin{array}{lll}
s^3: & 1 & 8 \\
s^2: & 5 & 6 + K \\
s^1: & (34 - K)/5 & \\
s^0: & 6 + K &
\end{array}
$$

from which for stability (that is, no sign changes in the first column) $K < 34$; therefore when $K = 34$, the particular locus crosses the imaginary axis at values of s given by

$$5s^2 + 40 = 0$$

that is

$$s = \pm j\sqrt{8}$$

Rule 11 Following application of rules 1 to 10 a good indication of the form of the root-locus diagram is obtained and the angle criterion can then be used to determine whether or not particular points in the complex plane lie on the loci. In this manner the full root-locus diagram may be obtained. Subsequently, the magnitude criterion can be used at any point on the root-locus to determine the value of K at that point

$$K = \frac{\text{product of the distances from the point to the poles}}{\text{product of the distances from the point to the zeros}} \quad (6.29)$$

and in the case where there are no zeros present the denominator is taken to be equal to unity.

Rule 12 In particular cases, and to assist rule 11, it may be possible to use the relationships between the coefficients of the characteristic equation and the sums and products of its roots. For example, from equation 6.18 the characteristic equation may be written as

$$a(s) + Kb(s) = 0 \quad (6.30)$$

that is

$$s^n + a_1 s^{n-1} + a_2 s^{n-2} + \ldots + a_n + K(s^m + b_1 s^{m-1} + \ldots + b_m) = 0$$

If the rank of the system is two or more, the coefficient of the s^{n-1} term is a_1 and is independent of K; this means that the sum of the roots and hence the sum of the poles remains constant at corresponding points throughout the locus.

Example 6.1
Determine the complete root-locus diagram for $0 \leq K \leq \infty$ for the closed-loop system of figure 6.3 with open-loop transfer function

$$G(s) = \frac{1}{(s+3)(s^2+2s+2)}$$

By rule 2 the number of poles of $G(s)$ is three, and therefore there are 3 loci.
By rule 3 the starting points of the loci are at $s = -3, s = -1 \pm j1$.
By rule 4 the finishing points are at infinity.
By rule 5 one branch of the root-locus lies between -3 and $-\infty$ along the real axis.
By rule 6 the rank is three, and therefore the asymptotes make angles of $-60°$, $-180°$, and $60°$ with the real axis.
By rule 7 the asymptotes intersect at $s = -5/3 = -1.66$.

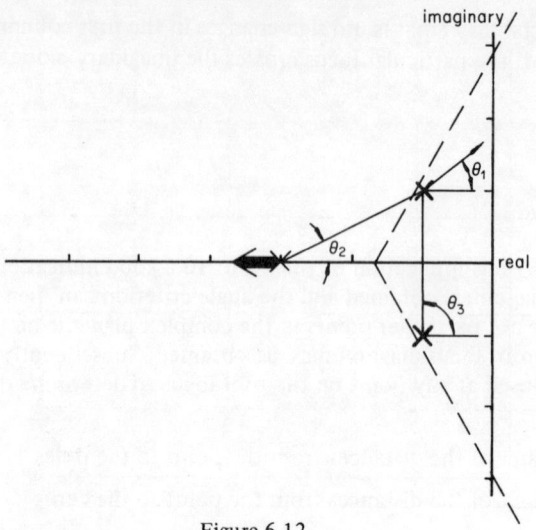

Figure 6.12

Figure 6.12 illustrates the results of applying rules 2 to 7.

By rule 9, θ_1, the angle of departure from the complex pole at $s = -1 + j1$, is given by

$$-\theta_1 - \theta_2 - \theta_3 = (2n + 1)\pi$$

that is

$$-\theta_1 - 26.6 - 90 = (2n + 1)\pi$$

and therefore

$$\theta_1 = 63.4°$$

By rule 10, the loci cross the imaginary axis at $s = \pm j\sqrt{8}$ when $K = 34$.

By rules 11 and 12 since the rank of the system is 3, the sum of the poles remains constant at -5, and so at the point when the two complex branches cross the imaginary axis, that is, at $s = \pm j\sqrt{8}$ the third pole is at $s = -5$ on the real branch of the locus.

The full locus is shown in figure 6.13; the values of the gain K obtained using rule 11 are indicated. From the diagram, for $K = 10$, the closed-loop-poles are $s = -4, s = -0.5 \pm j1.94$ and hence the closed-loop transfer function $G'(s)$ is given by

$$G'(s) = K \frac{\text{open-loop zeros}}{\text{closed-loop poles}} \qquad (6.31)$$

$$= 10 \left[\frac{1}{(s + 4)(s + 0.5 + j1.94)(s + 0.5 - j1.94)} \right]$$

$$= \frac{10}{(s + 4)(s^2 + s + 4.0)}$$

The Root-locus Method

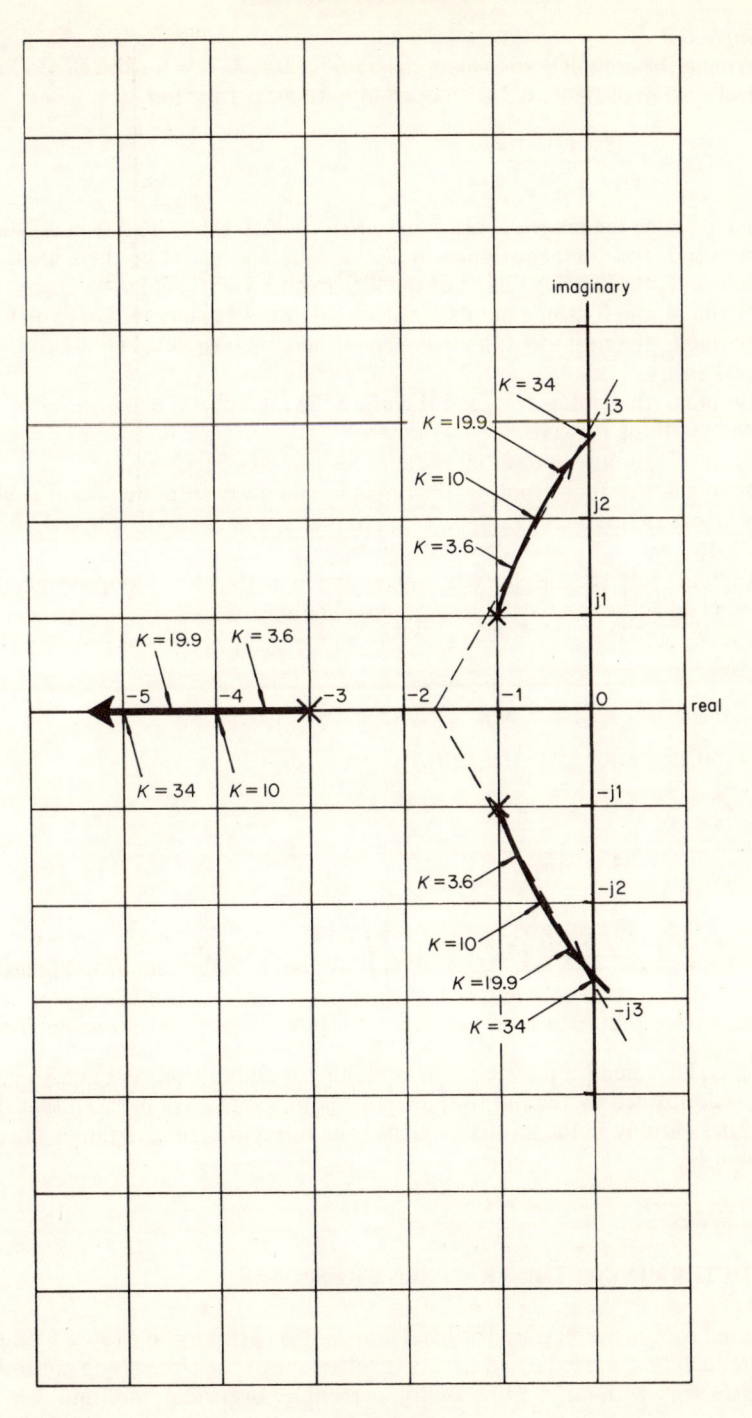

Figure 6.13

Example 6.2
Determine the complete root-locus diagram for $0 \leqslant K \leqslant \infty$ for the closed-loop control system of figure 6.3 with open-loop transfer function

$$G(s) = \frac{(s+1)(s+2)}{s^2(s+3)(s+4)}$$

Also determine the range of variation of K for which the system remains stable.

By rule 2, the number of poles of $G(s)$ is four, and therefore there are four loci.
By rule 3, the starting points of the loci are at $s = 0$ (twice), $s = -3$, $s = -4$.
By rule 4, the finishing points are at $s = -1$, $s = -2$ and there are two at infinity.
By rule 5, the root-locus lies between -1 and -2, and between -3 and -4 along the real axis.
By rule 6, the rank is two, and therefore the asymptotes make angles of $-90°$ and $90°$ with the real axis.
By rule 7, the asymptotes intersect at $s = [-7-(-3)]/2 = -2$.
By rule 8, the double pole at the origin breaks away from the origin at $90°$ to the real axis. Also, there is a departure point at $s = -3.44$ and an arrival point at $s = -1.46$.

Applying rule 10, the characteristic equation of the closed-loop system is

$$s^2(s+3)(s+4) + K(s+1)(s+2) = 0$$

that is

$$s^4 + 7s^3 + (12+K)s^2 + 3Ks + 2K = 0$$

The Routh criterion gives the array

s^4	1	$12 + K$	$2K$
s^3	7	$3K$	
s^2	$\dfrac{84 + 4K}{7}$	$2K$	
s^1	$\left[\left(\dfrac{84+4K}{7}\right)3K - 14K\right] / \left(\dfrac{84+4K}{7}\right)$		
s^0	$2K$		

Thus the closed-loop system will be stable for all positive values of K and this will be confirmed by the full root-locus diagram where it will be seen that the root-locus lies entirely in the left-half s-plane. The full root-locus diagram is shown in figure 6.14.

6.4 DETERMINING THE TRANSIENT RESPONSE

Once the root-locus diagram has been drawn, it is necessary to choose a value of the gain K such that the expected transient behaviour of the closed-loop control system is satisfactory with respect to stability, percentage overshoot, etc., and then to

The Root-locus Method

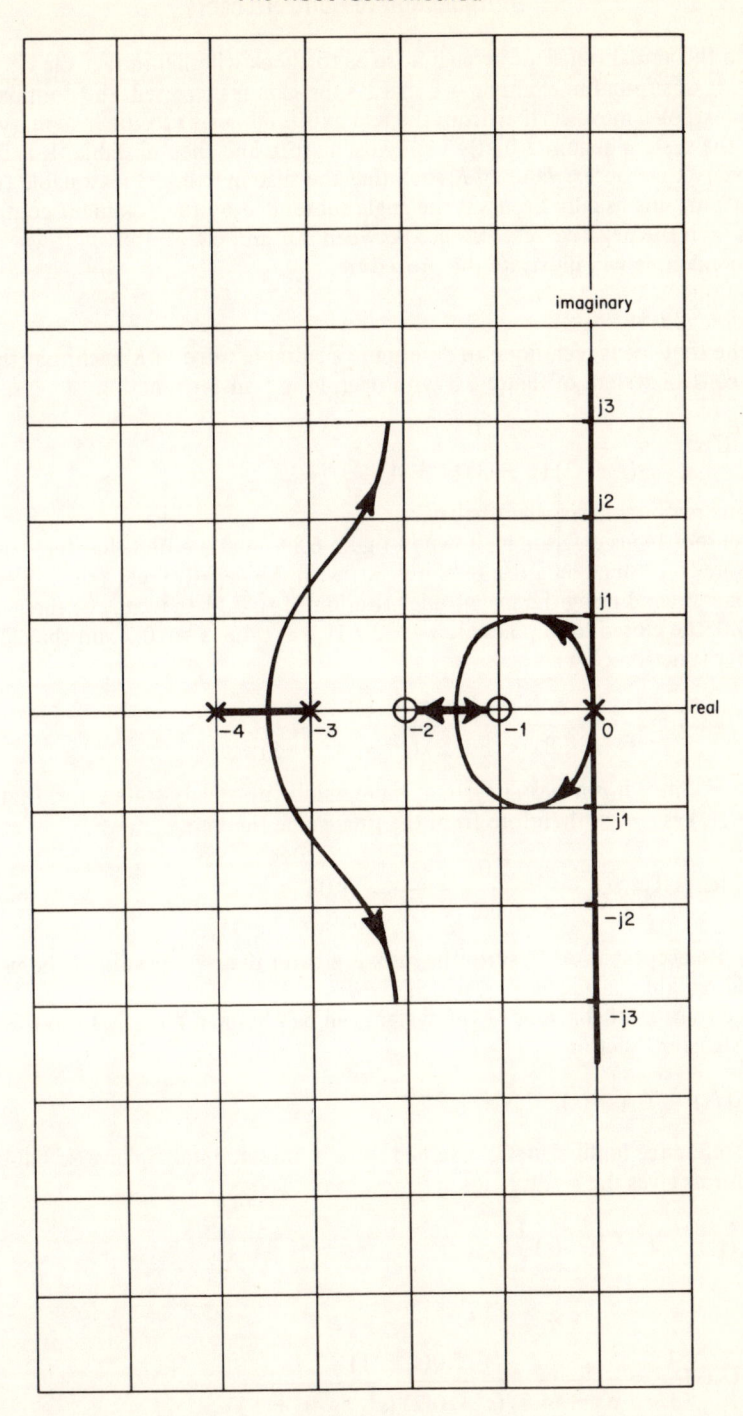

Figure 6.14

obtain the actual transient behaviour so as to check whether in fact the choice is good. It was found in example 6.1 that, as the gain is increased, the dominant complex poles move further from the real axis (and nearer to the imaginary axis) until the system becomes firstly highly oscillatory and then unstable. It is thus necessary to choose a value of K such that the system exhibits reasonable transient behaviour; this usually occurs if the angle subtended by the dominant complex poles with the negative real axis lies between 30° and 60°.

An example will illustrate the procedure.

Example 6.3

Use the root-locus technique to determine a suitable value of K such that the closed-loop control system of figure 6.3 with open-loop transfer function

$$G(s) = \frac{s+1}{s(s+2)(s+3)(s+4)}$$

exhibits reasonable transient behaviour.

The root-locus diagram is shown in figure 6.15. Assume first that the complex poles are to subtend an angle of about 30° with the negative real axis. Thus from the diagram and using the magnitude criterion of rule 11 the value of the gain is $K = 5.0$, the closed-loop poles are $s = -2 \pm j1$, $s = -4.8$, $s = -0.2$ and the closed-loop transfer function $G'(s)$ is

$$G'(s) \approx \frac{5.0(s+1)}{(s+4.8)(s+0.2)(s^2+4s+5)}$$

Check Since it is a class-1 system, there will be no steady-state error, that is, $c(t) = r(t)$ as $t \to \infty$; therefore from the final-value theorem

$$[G'(s)]_{s=0} = \frac{5.0}{4.8 \times 0.2 \times 5.0} = 1.04 \approx 1.0$$

which is acceptable. Also, since the rank is greater than 1, the sum of the poles should remain constant at -9, which is so.

The transient behaviour of the system can be obtained directly from the closed-loop transfer function

$$c(s) = G'(s)r(s) \tag{6.32}$$

if the reference input signal is assumed to be a unit step, taking inverse Laplace transforms gives the result

$$c(t) = \mathcal{L}^{-1}\left[G'(s)\frac{1}{s}\right] \tag{6.33}$$

Now

$$G'(s)\frac{1}{s} = \frac{5.0(s+1)}{s(s+4.8)(s+0.2)(s^2+4s+5)}$$

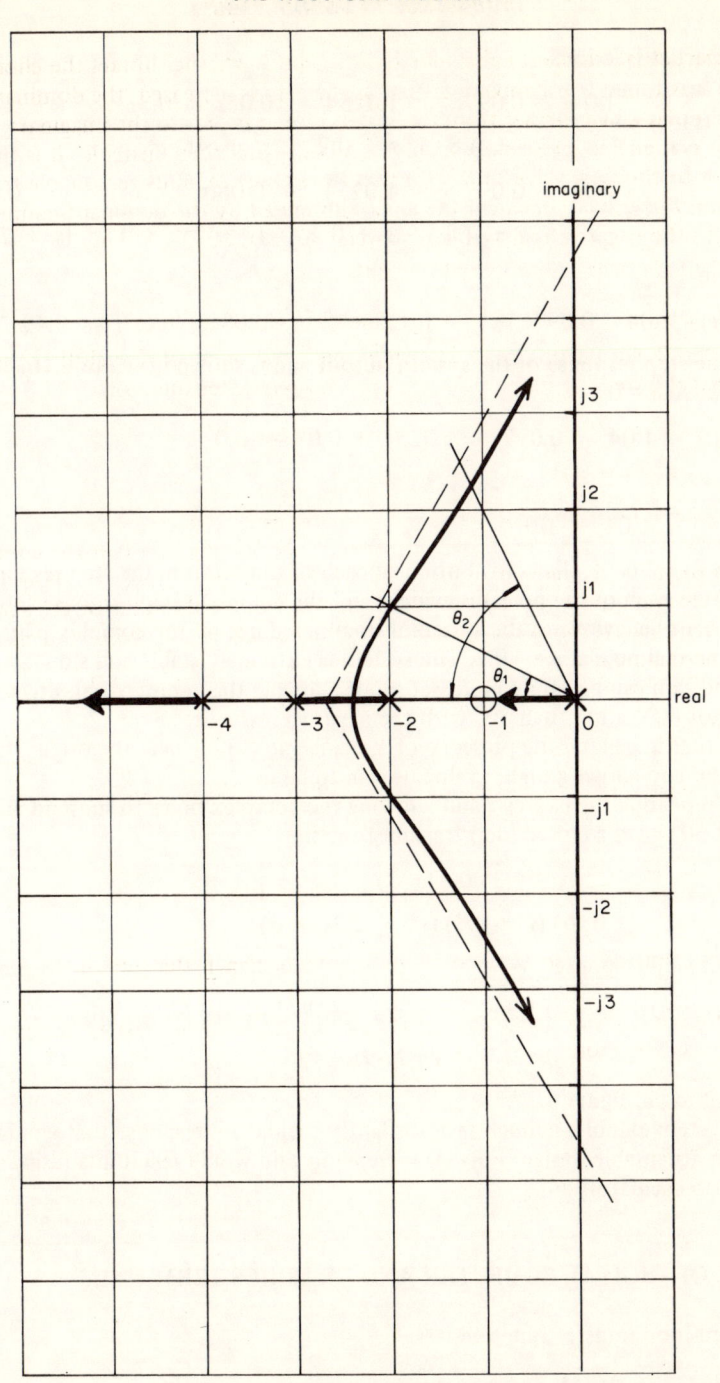

Figure 6.15

and by partial fractions

$$G'(s)\frac{1}{s} = \frac{1.04}{s} - \frac{0.097}{s+4.8} - \frac{1.0254}{s+0.2} + \frac{0.08s - 0.33}{s^2 + 4s + 5}$$

$$= \frac{1.04}{s} - \frac{0.097}{s+4.8} - \frac{1.0254}{s+0.2} + \frac{0.08(s+2)}{(s+2)^2 + 1} - \frac{0.49}{(s+2)^2 + 1}$$

Therefore

$$c(t) = 1.04 - 0.097e^{-4.8t} - 1.0254e^{-0.2t} + 0.08e^{-2t}\cos t - 0.49e^{-2t}\sin t$$

This is the step response of the system output and is shown in figure 6.16.

Check. At $t = 0$

$$c(0) = 1.04 - 0.097 - 1.0254 + 0.08 \approx 0.0$$

At $t = \infty$

$$c(\infty) = 1.04 \approx 1.0$$

From figure 6.16, the contribution of each of the terms in the step response (and in fact from each of the poles) is evident and the complete step response shows that the transient behaviour of the system is dominated not by the complex pair of poles, but by the real pole at $s = -0.2$. The system is extremely stable and slow acting and hence the gain can be safely increased, thus allowing the complex poles to subtend an angle of 60° rather than 30° with the negative real axis.

Note that in general the presence of a real negative pole near the origin stabilises the system and allows a higher value of gain to be used.

Thus repeating the exercise and allowing the complex poles to subtend an angle of about 60° gives a closed-loop transfer function

$$G'(s) \approx \frac{28.0(s+1)}{s(s+6)(s+0.8)(s^2 + 2.2s + 6)}$$

and a unit step response

$$c(t) = 0.975 - 0.159e^{-6t} - 0.276e^{-0.8t} - 0.53e^{-1.1t}\cos 2.19t$$
$$- 0.82e^{-1.1t}\sin 2.19t$$

which is given in figure 6.17.

The system exhibits a much faster, slightly oscillatory response that would constitute an acceptable design. Further increase in gain would result in a response that was far too oscillatory.

6.5 USE OF DIGITAL COMPUTER AND GRAPHICS TERMINAL

The closed-loop transfer function is

$$G'(s) = \frac{Kb(s)}{a(s) + Kb(s)} \tag{6.34}$$

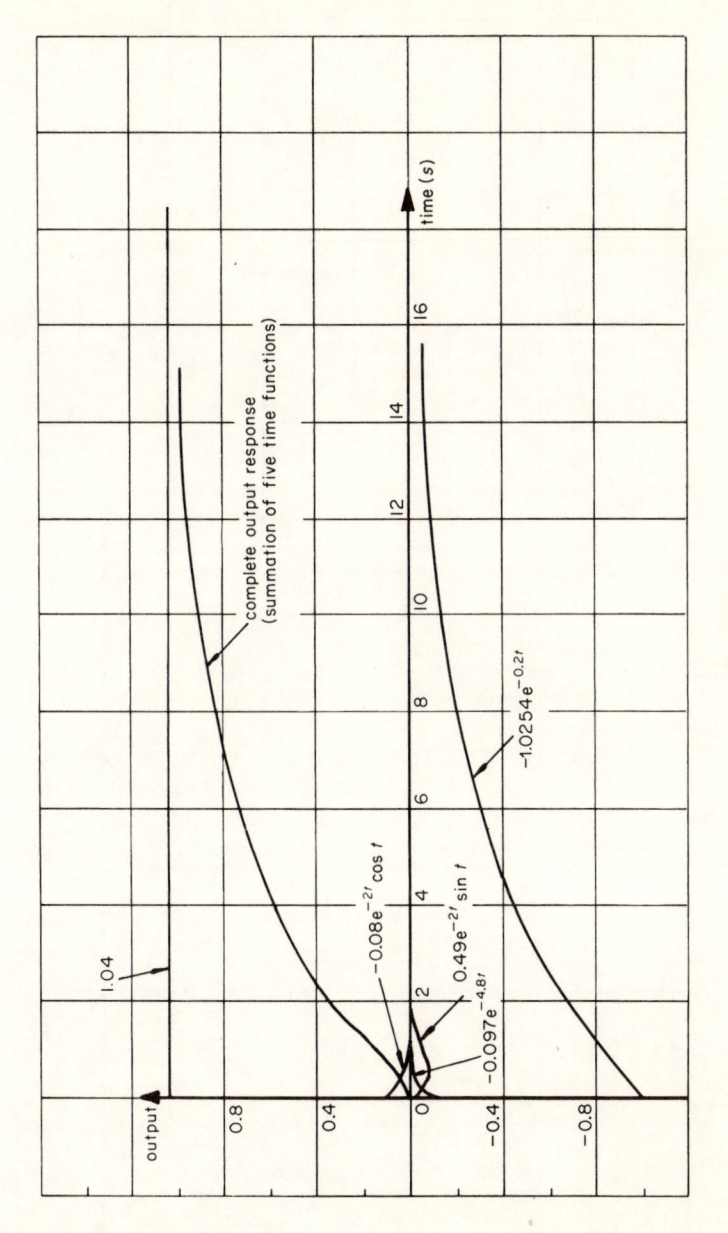

Figure 6.16

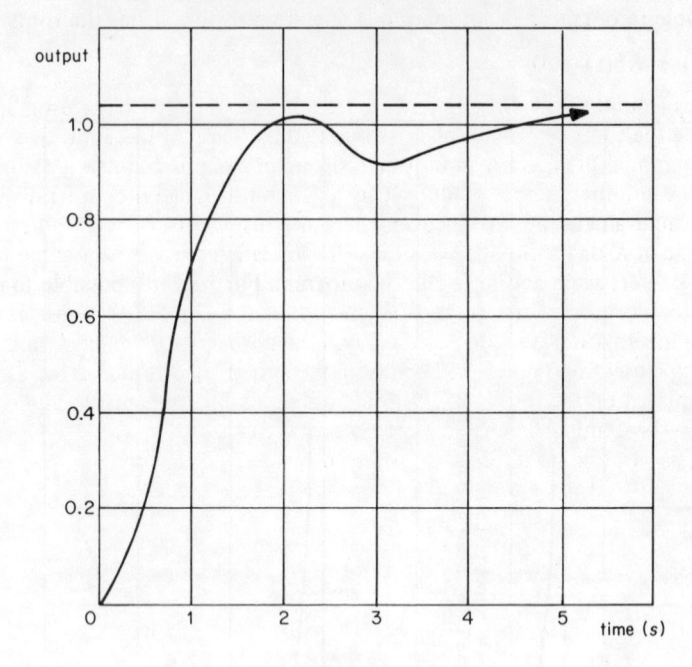

Figure 6.17

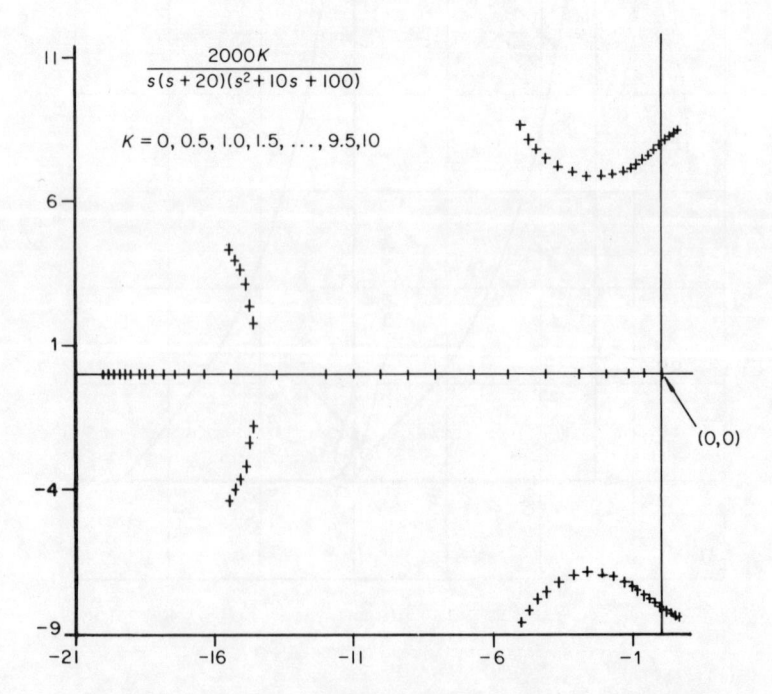

$$\frac{2000K}{s(s+20)(s^2+10s+100)}$$

$K = 0, 0.5, 1.0, 1.5, \ldots, 9.5, 10$

Figure 6.18

The Root-locus Method

It is possible to obtain the root-locus diagram—that is, the loci of the roots of

$$a(s) + Kb(s) = 0 \tag{6.35}$$

as K varies from 0 to ∞—directly using a digital computer that has a program for calculating the roots of polynomials. If the digital computer has an interface with a graphics terminal then, given an input consisting of the polynomials $a(s)$ and $b(s)$, the range K and the number of steps N into which this range should be divided, the program can evaluate the root-locus diagram and display it on the graphics terminal. If the range of K or the number of steps N is not satisfactory it is possible to input these parameters again and to re-run the program. Further, it is possible to identify the required position of the dominant poles by using a cursor on the display and to input this information; the program will then identify the gain value K and the corresponding closed-loop poles, will compute the system step response for a given time T and display it on the terminal. If this is not acceptable, the cursor can be used to re-identify the dominant poles and the program re-run.

The type of program outlined above is known as a *conversational program* and an example of the output is shown in figures 6.18 and 6.19.

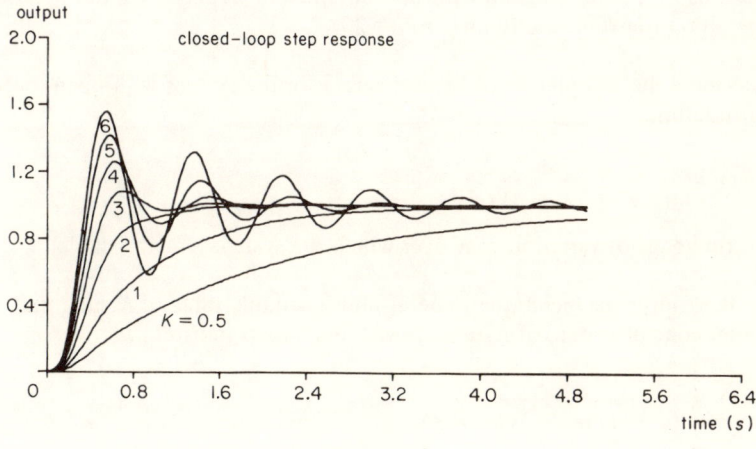

Figure 6.19

REFERENCES

Dransfield, P., and Haber, D. F., *Introducing Root Locus* (Cambridge University Press, 1973).
Evans, W. R., Graphical Analysis of Control Systems, *Trans., A.I.E.E.* 67 (1948) 547–51.
Taylor, P. L., *Servomechanisms* (Longmans, London, 1960).

PROBLEMS

6.1 Draw the pole–zero configuration for the system transfer function

$$G(s) = \frac{s + 3.0}{(s + 3.2)(s + 1.0)(s^2 + 4s + 13)}$$

Discuss the contribution of each pole to the system impulse response and calculate the value of $G(s)$ at $s = -2.5$ and $s = 1 + 3j$.

6.2 Draw the pole–zero configuration for the system transfer function

$$G(s) = \frac{s + 2.0}{(s + 2.1)(s + 0.5)(s + 4.0)}$$

and discuss the contribution of each pole to the system impulse response. Use partial fractions and inverse Laplace transforms to obtain this impulse response for $0 \leqslant t \leqslant 4$ s.

6.3 Determine the complete root-locus diagram for $0 \leqslant K \leqslant \infty$ for the closed-loop system of figure 6.3 with open-loop transfer function

$$G(s) = \frac{1}{(s + 1)(s^2 + 6s + 13)}$$

Compare this root-locus diagram with that obtained in example 6.1 and determine the closed-loop transfer function when $K = 20$.

6.4 Determine the complete root-locus diagram for the system with open-loop transfer function

$$G(s) = \frac{s + 1}{s(s + 2)(s + 4)(s^2 + 2s + 5)}$$

and find the range of variation of K over which the system remains stable.

6.5 Use the root-locus technique to determine a suitable value of K such that the closed-loop control system of figure 6.3 with open-loop transfer function

$$G(s) = \frac{1}{(s + 1)(s + 2)(s + 3)}$$

exhibits reasonable transient behaviour.

7 Frequency-response Methods

An important branch of control theory is concerned with the frequency response of a system. This approach to the investigation of system performance provides an alternative to the transient-response and root-locus methods and is favoured by many control engineers, especially those with an electrical-engineering background. The frequency-response method considers system behaviour due to sinusoidal forcing; the information obtained can be used to determine closed-loop stability from the open-loop frequency response in a manner analogous to the root-locus method.

Thus, when studying transient response, attention is directed to the decay of that response with time; the root-locus method is concerned with the position of the roots in the s-plane; and with frequency-response methods it is possible to infer these positions using certain theorems.

A further (and important) reason for the use of the frequency response is that in many cases the transfer function of the system to be controlled cannot be obtained from analytical considerations as outlined in chapter 2, but may be obtained by experimentation. A suitable method is to excite the system with a range of sinusoidal signals—usually of similar amplitude but varying frequency—and to observe the resulting output from the system in the form of the amplitude and phase shift. Using this information (which is in fact one method of representing frequency response) it is possible to design an appropriate control system.

The present chapter considers means of obtaining the frequency response of linear systems and various graphical representations of this response (the frequency-domain method, like the root-locus method, is a graphical approach to control-system design), including polar plots (Nyquist diagrams), amplitude-frequency and phase-frequency plots (Bode diagrams) and amplitude-phase plots (Nichols diagrams). It is shown how the stability (both absolute and relative) of the closed-loop system can be obtained by using the polar plot of the open-loop system together with Nyquist's stability criterion. The results are interpreted with reference to the other graphical representations and it is a matter of personal choice which approach is used.

The chapter concludes with the interpretation of these methods in order to obtain insight into closed-loop performance and transient behaviour.

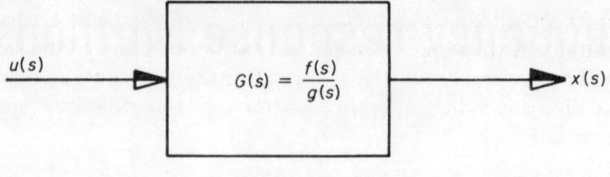

7.1 DERIVATION OF THE FREQUENCY RESPONSE OF A SYSTEM

Consider the open-loop system of figure 7.1, where the input signal $u(t)$ is a sinusoid of unity amplitude and frequency ω, that is, $u(t) = \sin \omega t$; the resulting output is obtained from the transformed relationship

$$x(s) = G(s)u(s) = \frac{f(s)}{g(s)} u(s) = \frac{f(s)}{g(s)} \frac{\omega}{s^2 + \omega^2} \qquad (7.1)$$

where $\omega/(s^2 + \omega^2)$ is the Laplace transform of the input signal. This relationship may be written as the sum of two terms

$$x(s) = \frac{k(s)}{g(s)} + \frac{As + B}{s^2 + \omega^2} \qquad (7.2)$$

$$= \frac{k(s)(s^2 + \omega^2) + g(s)(As + B)}{g(s)(s^2 + \omega^2)} \qquad (7.3)$$

Comparing the numerators of equations 7.1 and 7.3 gives

$$k(s)(s^2 + \omega^2) + g(s)(As + B) \equiv f(s)\omega \qquad (7.4)$$

and, since this is an identity, it holds for all values of s. Put $s = j\omega$; assuming that $g(s)$ contains no $(s^2 + \omega^2)$ terms

$$g(j\omega)(Aj\omega + B) = f(j\omega)\omega \qquad (7.5)$$

that is

$$Aj\omega + B = f(j\omega)\omega/g(j\omega) = G(j\omega)\omega \qquad (7.6)$$

and $G(j\omega)$ is obtained directly from the open-loop transfer function $G(s)$ merely by replacing s by $j\omega$.

Equating real and imaginary parts of equation 7.6 gives

$$B = R[G(j\omega)]\omega \qquad A = I[G(j\omega)] \qquad (7.7)$$

and substituting these values into equation 7.2

$$x(s) = \frac{k(s)}{g(s)} + \frac{Is + R\omega}{s^2 + \omega^2} \qquad (7.8)$$

The first term of this equation represents the transient part of the response and in partial-fraction form will give rise to terms with denominators of the form $(s + a)$

Frequency-response Methods

and $(s^2 + bs + c)$. Provided $G(s)$ represents a *stable* system, these terms will die away exponentially as time increases. The second term is due to the sinusoidal forcing and will be present at all times, even after the transients associated with the first term have died away; hence, from equation 7.8, the ultimate response resulting from the forcing will be

$$x(s) = \frac{Is + R\omega}{s^2 + \omega^2} \tag{7.9}$$

Therefore

$$x(t) = \mathcal{L}^{-1}\frac{Is + R\omega}{s^2 + \omega^2} = I\mathcal{L}^{-1}\frac{s + (R/I)\omega}{s^2 + \omega^2}$$

$$= \frac{I}{\omega}\left(\frac{R^2}{I^2}\omega^2 + \omega^2\right)^{\frac{1}{2}} \sin\left(\omega t + \tan^{-1}\frac{\omega I}{\omega R}\right) \tag{7.10}$$

using the table of inverse Laplace transforms (see appendix). Therefore

$$x(t) = (R^2 + I^2)^{\frac{1}{2}} \sin\left(\omega t + \tan^{-1}\frac{I}{R}\right) \tag{7.11}$$

$$= |G(j\omega)| \sin(\omega t + \phi) \tag{7.12}$$

where

$$\phi = \arg G(j\omega) \tag{7.13}$$

Equation 7.12 represents the frequency response of the system with transfer function $G(s)$. It is evident that, if the system is excited by a sinusoid of unity amplitude, the system output (once the transients have died away) is also a sinusoid of equal frequency with amplitude $|G(j\omega)|$ and phase angle $\phi = \arg G(j\omega)$; this situation is shown in figure 7.2.

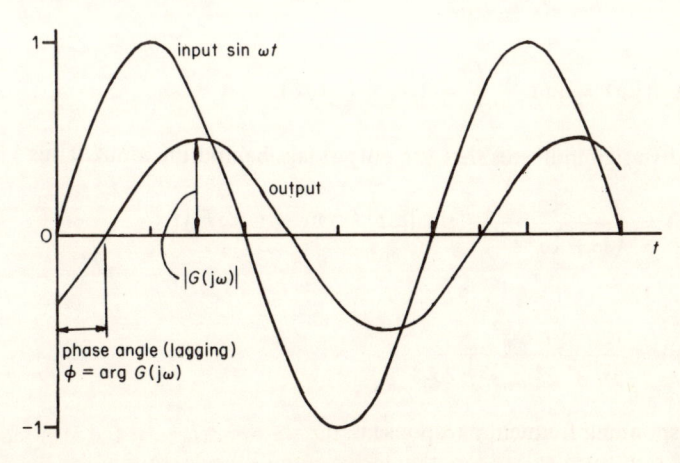

Figure 7.2

Introduction to Control Theory

Thus, all that is required to obtain the system frequency-response function given the system transfer function is to replace s by $j\omega$ and then calculate the resulting modulus and argument.

Example 7.1

Obtain the frequency-response functions of the systems with transfer functions $1/(1 + sT)$ and $\omega_n^2/(s^2 + 2\xi\omega_n s + \omega_n^2)$.

(i) If

$$G(s) = \frac{1}{1 + sT}$$

the corresponding frequency response is

$$G(j\omega) = \frac{1}{1 + j\omega T} = \frac{1 - j\omega T}{1 + \omega^2 T^2} = R(\omega) + jI(\omega)$$

Therefore

$$R(\omega) = \frac{1}{1 + \omega^2 T^2}$$

$$I(\omega) = -\frac{\omega T}{1 + \omega^2 T^2}$$

Hence

$$|G(j\omega)| = (R^2 + I^2)^{\frac{1}{2}} = \left[\left(\frac{1}{1 + \omega^2 T^2}\right)^2 + \left(\frac{\omega T}{1 + \omega^2 T^2}\right)^2\right]^{\frac{1}{2}}$$

$$= \left(\frac{1}{1 + \omega^2 T^2}\right)^{\frac{1}{2}}$$

and

$$\arg G(j\omega) = \tan^{-1}\frac{I}{R} = \tan^{-1}(-\omega T)$$

The negative sign indicates that the output lags behind the input. Thus

$$x(t) = \left(\frac{1}{1 + \omega^2 T^2}\right)^{\frac{1}{2}} \sin\left[\omega t + \tan^{-1}(-\omega T)\right] \tag{7.14}$$

(ii) If

$$G(s) = \frac{\omega_n^2}{s^2 + 2\xi\omega_n s + \omega_n^2}$$

the corresponding frequency response is

Frequency-response Methods

$$G(j\omega) = \frac{\omega_n^2}{(j\omega)^2 + 2\xi\omega_n(j\omega) + \omega_n^2} = \frac{\omega_n^2}{(\omega_n^2 - \omega^2) + j2\xi\omega_n\omega}$$

$$= \frac{\omega_n^2[(\omega_n^2 - \omega^2) - j2\xi\omega_n\omega]}{(\omega_n^2 - \omega^2)^2 + (2\xi\omega_n\omega)^2}$$

Now define u as the ratio between the applied (input) sinusoidal frequency and the undamped natural frequency

$$u = \frac{\omega}{\omega_n} \qquad (7.15)$$

Then

$$G(j\omega) = \frac{(1 - u^2) - j2\xi u}{(1 - u^2)^2 + (2\xi u)^2}$$

and

$$\left. \begin{array}{l} |G(j\omega)| = \left[\dfrac{1}{(1 - u^2)^2 + (2\xi u)^2} \right]^{\frac{1}{2}} \\[2mm] \arg G(j\omega) = \tan^{-1}\left(-\dfrac{2\xi u}{1 - u^2} \right) \end{array} \right\} \qquad (7.16)$$

Again the output lags behind the input. Equations 7.14 and 7.16 represent important fundamental results and will be referred to throughout this chapter.

For a system whose transfer function is represented by a product of several simpler transfer functions

$$G(s) = G_1(s)G_2(s) \ldots G_n(s) \qquad (7.17)$$

the over-all frequency-response function is obtained simply by forming the product of the amplitude functions of each of the individual terms and the sum of the phase functions. Thus

$$|G(j\omega)| = |G_1(j\omega)| \, |G_2(j\omega)| \ldots |G_n(j\omega)| \qquad (7.18)$$

and

$$\arg G(j\omega) = \arg G_1(j\omega) + \arg G_2(j\omega) + \ldots + \arg G_n(j\omega) \qquad (7.19)$$

Example 7.2
Obtain the frequency-response function of the system with transfer function

$$G(s) = \frac{1}{(s + 1)(2s + 1)(3s + 1)}$$

Using the results of equations 7.14, 7.18 and 7.19

$$|G(j\omega)| = \left(\frac{1}{1 + \omega^2}\right)^{\frac{1}{2}} \left(\frac{1}{1 + 4\omega^2}\right)^{\frac{1}{2}} \left(\frac{1}{1 + 9\omega^2}\right)^{\frac{1}{2}}$$

and

$$\arg G(j\omega) = -\tan^{-1} \omega - \tan^{-1} 2\omega - \tan^{-1} 3\omega$$

and hence $x(t)$ is found.

It is evident from equation 7.12 that, as ω is increased from 0 to ∞, the amplitude and phase characteristics of the system will change, thus giving the full frequency response of the system. These characteristics can best be represented and interpreted graphically and this is done in subsequent sections: the object of doing so is to obtain information on system stability.

Before proceeding, it is worthwhile to consider the closed-loop behaviour of a system that is excited by a sinusoid, so as to establish the possibility of demonstrating that a system is unstable. Consider the closed-loop system of figure 7.3,

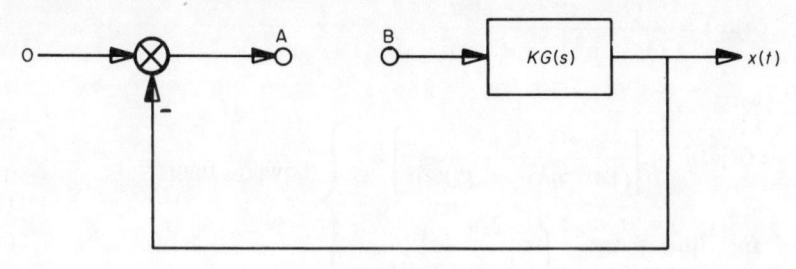

Figure 7.3

where a break has been made between points A and B. Suppose that at point B the signal is $\sin \omega t$; then the output signal can be represented by $R \sin(\omega t + \phi)$ and the signal at point A will be $-R \sin(\omega t + \phi)$.

Case 1: $\phi = 180°$, $R < 1$. At a particular frequency suppose the phase angle to be $180°$ and the amplitude R to be less than 1; then $x_B = \sin \omega t$ and $x_A = R \sin \omega t$ and the signal will die out, indicating a stable system.

Case 2: $\phi = 180°$, $R = 1$. In this case $x_B = \sin \omega t$ and $x_A = \sin \omega t$; this signal will travel around the loop indefinitely, indicating that the system is on the threshold of instability.

Case 3: $\phi = 180°$, $R > 1$. The magnitude of the signal at A will be greater than that at B and will grow indefinitely, indicating instability.

Thus, it has been shown heuristically that knowledge of the gain and phase characteristics of the open-loop system can indicate its behaviour in the closed loop.

Interpretation in the s-plane

It was shown in the previous chapter that the function $G(s)$ could be expressed as

$$G(s) = M(s)\exp[j\gamma(s)] = |G(s)|\arg G(s) = |G(s)| \angle \gamma(s) \qquad (7.20)$$

where $|G(s)|$ is the ratio between the product of the distances from the given point s to the zeros and the product of the distances from s to the poles; $\arg G(s)$ is the difference between the sum of the angles subtended by the zeros and the sum of the angles subtended by the poles.

Frequency-response Methods

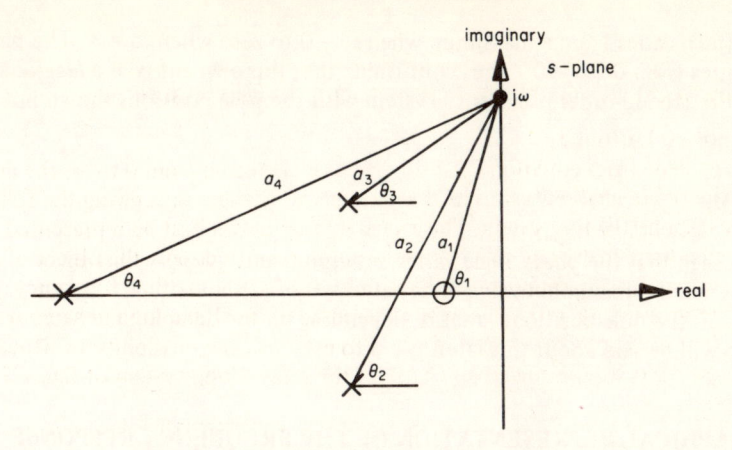

Figure 7.4

Thus

$$G(j\omega) = |G(j\omega)| \arg G(j\omega) \tag{7.21}$$

Hence for the particular pole-zero configuration shown in figure 7.4

$$G(j\omega) = a_1/a_2 a_3 a_4$$
$$\arg G(j\omega) = \theta_1 - \theta_2 - \theta_3 - \theta_4 \tag{7.22}$$

As the point moves from the origin along the positive imaginary axis ω will change from 0 to ∞, and so the quantities in equation 7.22 will change. Thus, as far as the frequency response is concerned the motion of the point s is restricted to the imaginary axis only; this is in contrast with the root-locus method, where s could assume any value within the complete s-plane, that is $s = \sigma + j\omega$.

When $G(s) = 1/(1 + sT)$ (a first-order lag) it follows from figure 7.5a that

$$G(j\omega) = |G(j\omega)| \arg G(j\omega) = \frac{1}{a} \angle -\theta \tag{7.23}$$

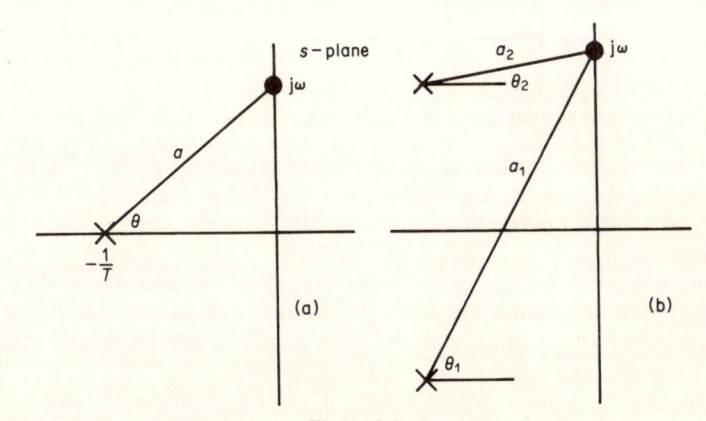

Figure 7.5

and $|G(j\omega)|$ varies from a maximum when $\omega = 0$ to zero when $\omega = \infty$. The phase angle varies from 0 to $-90°$, thus confirming that this element is of a *laggy* nature.

For the second-order oscillatory system with the pole positions shown in figure 7.5b

$$G(j\omega) = \frac{1}{a_1 a_2} \angle -\theta_1 - \theta_2 \qquad (7.24)$$

and it is seen that the phase angle varies between 0 and $-180°$ whilst the modulus can experience a maximum value either at $\omega = 0$ or at some other frequency, depending on the pole positions—that is, depending on the damping factor ξ.

More will be said about this later.

7.2 GRAPHICAL REPRESENTATION OF THE FREQUENCY-RESPONSE FUNCTION

The frequency-response function $G(j\omega) = |G(j\omega)| \arg G(j\omega)$ can be represented graphically by plotting both $|G(j\omega)|$ and $\arg G(j\omega)$ as functions of ω for $0 \leqslant \omega < \infty$ in various forms, including the following.

(1) In the *polar plot* the point $G(j\omega_1)$ is obtained by drawing a line of magnitude $|G(j\omega_1)|$ at an angle ϕ equal to $\arg G(j\omega_1)$ with the positive real axis, counterclockwise being taken as positive; this is shown in figure 7.6 and is a *vector plot*.

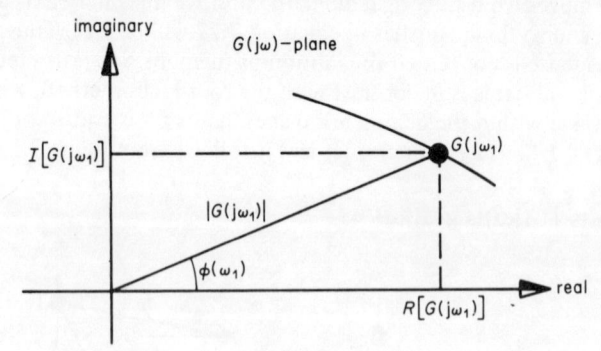

Figure 7.6

This is obviously equivalent to obtaining the real and imaginary parts of $G(j\omega)$ and plotting them on an Argand diagram but it can be seen that the polar plot is more easily (and conveniently) obtained. The polar plot is known as a *Nyquist diagram*.

(2) The amplitude–frequency and phase–frequency representation consists of *two* graphs on a common frequency (horizontal) axis, as shown in figure 7.7. It is usual to use a logarithmic scale for both amplitude and frequency and a linear scale for phase: the result is called a *Bode diagram*.

(3) The amplitude–phase representation is a single graph incorporating a logarithmic scale for amplitude and a linear scale for phase, known as a *Nichols diagram* (figure 7.8).

Frequency-response Methods

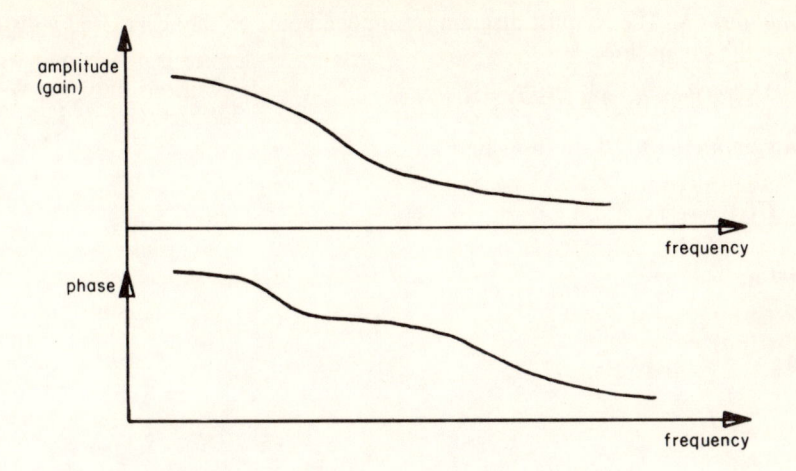

[Figure 7.7: amplitude (gain) vs frequency, and phase vs frequency curves]

Figure 7.7

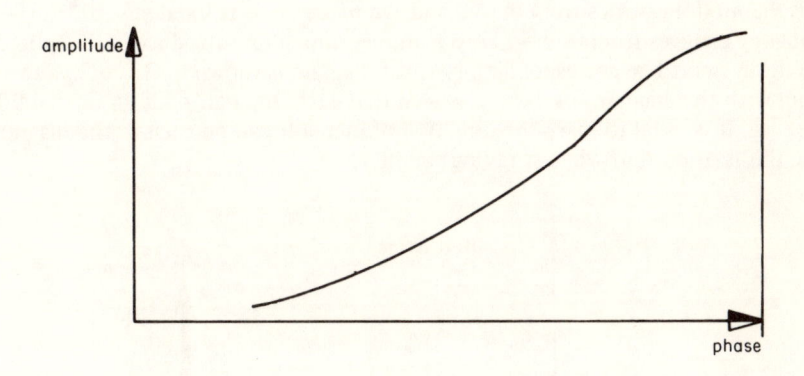

[Figure 7.8: amplitude vs phase curve]

Figure 7.8

Polar Plot or Nyquist Diagram

Usually the open-loop-system transfer function $G(s)$ is represented as a product of simple factors in the general form

$$G(s) = \frac{K(a_1 s + 1) \ldots (b_1 s^2 + c_1 s + 1) \ldots}{s^l (d_1 s + 1) \ldots (e_1 s^2 + f_1 s + 1) \ldots} \quad (7.25)$$

where the order of the numerator is less than the order of denominator. Note the difference between this and previous representations of these factors, the reason for this will become evident later. The relationships expressed in equations 7.18 and 7.19 mean that once the Nyquist diagrams of the simple factors

$$K, s^l (as + 1), (bs^2 + cs + 1)$$

have been established the diagrams of more complicated systems can be derived.

Introduction to Control Theory

Gain term K. The Nyquist diagram is simply a point on the real axis at a distance K from the origin since

$$|G(j\omega)| = K; \quad \arg G(j\omega) = 0 \tag{7.26}$$

Integrating term $1/s$. In this case

$$G(s) = \frac{1}{s}$$

Therefore

$$G(j\omega) = \frac{1}{j\omega}$$

and

$$|G(j\omega)| = \frac{1}{\omega}; \quad \arg G(j\omega) = \tan^{-1}(-\infty) = -90° \tag{7.27}$$

Thus the modulus varies from ∞ to 0 and the phase angle is fixed at −90° as the frequency changes from 0 to ∞. Thus an integrating term introduces a 90° phase *lag* into the system. The corresponding Nyquist diagram is shown in figure 7.9a together with the diagram for $1/s^2$; it is seen that each integrator adds a further 90° phase lag. It is usual to mark the direction of increasing ω on each of the diagrams since this is important when studying stability.

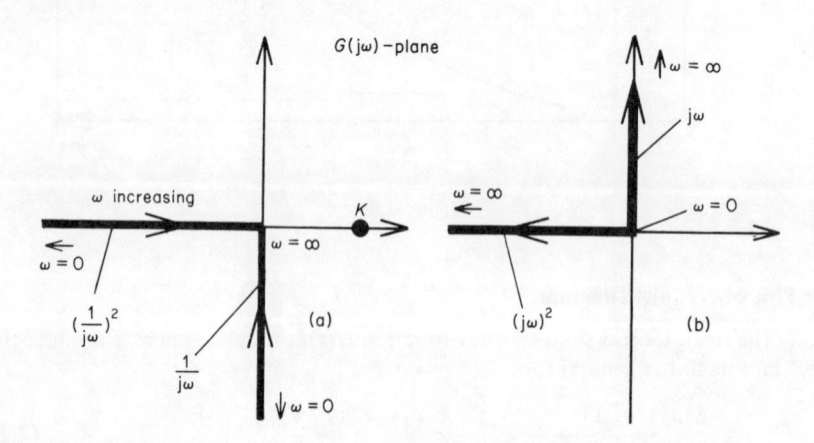

Figure 7.9

Differentiating term s. In this case

$$G(s) = s$$

and

$$G(j\omega) = j\omega$$

Frequency-response Methods

Therefore

$$|G(j\omega)| = \omega; \arg G(j\omega) = \tan^{-1}\infty = 90° \qquad (7.28)$$

This indicates that a differentiation introduces a 90° phase *lead* (or advance) into the system at all frequencies. The Nyquist diagrams for s and s^2 are shown in figure 7.9b.

First-order terms in denominator $1/(Ts + 1)$. These are referred to as first-order lag terms and from equation 7.14

$$|G(j\omega)| = \left(\frac{1}{1 + \omega^2 T^2}\right)^{\frac{1}{2}} \; ; \; \arg G(j\omega) = \tan^{-1}(-\omega T) \qquad (7.29)$$

hence, at zero frequency, the modulus is unity and the phase angle zero; at infinite frequency, the modulus is zero and the phase angle $-90°$. When the frequency $\omega = 1/T$, that is, $\omega T = 1$, the modulus is $1/\sqrt{2}$ and the phase angle $-45°$. Figure 7.10 shows this particular plot to be a semi-circular arc and indicates the method of construction of the plot using the gain and phase (as a vector) at each value of frequency.

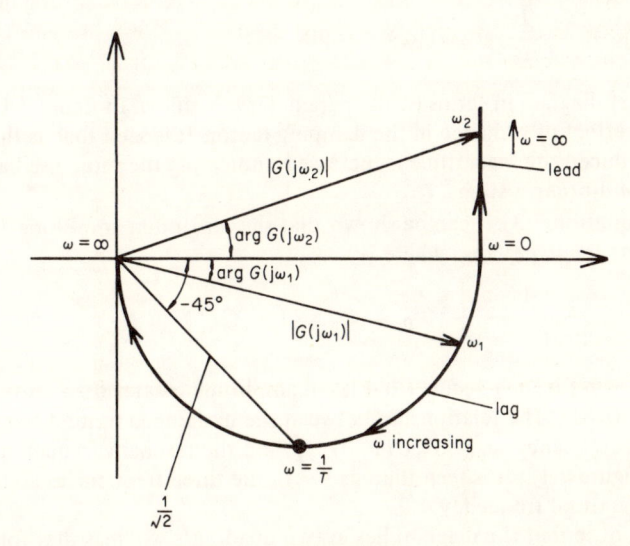

Figure 7.10

First-order term in numerator $Ts + 1$. These are referred to as first-order lead terms where

$$|G(j\omega)| = (1 + \omega^2 T^2)^{\frac{1}{2}}; \arg G(j\omega) = \tan^{-1} \omega T \qquad (7.30)$$

It is seen that the phase angle is always positive. Figure 7.10 shows the Nyquist diagram.

Second-order term in denominator $1/(T^2 s^2 + 2\xi Ts + 1)$. Let $T = 1/\omega_n$ and $u = \omega/\omega_n$; then the amplitude and argument are given by equation 7.16 and rewritten here for convenience as

$$G(j\omega) = \left[\frac{1}{(1-u^2)^2 + (2\xi u)^2}\right]^{\frac{1}{2}}$$

$$\arg G(j\omega) = \tan^{-1}\left[-\frac{2\xi u}{1-u^2}\right]$$
(7.31)

Thus at a given frequency ω, both quantities are functions of the undamped natural frequency ω_n and the damping factor ξ. For low-frequency input signals, the output follows the input closely but with a small lag, whilst for high-frequency inputs the output amplitude is small and lags the input by 180°. Some specific values are as follows: for $u = 0$

$$|G(j\omega)| = 1; \arg G(j\omega) = 0$$

for $u = \infty$

$$|G(j\omega)| = 0; \arg G(j\omega) = -180°$$

for $u = 1$ (that is, $\omega = \omega_n$)

$$|G(j\omega)| = \frac{1}{2\xi}; \arg G(j\omega) = -90°$$

The Nyquist diagram in terms of the parameter u is shown in figure 7.11 and illustrates the effect of a change in the damping factor. It is seen that as the damping factor is reduced the magnitude is increased, indicating the more oscillatory character of the second-order system.

From equation 7.31 it can be shown that the maximum amplitude M_p occurs at the *resonant frequency* ω_0, where

$$M_p = \frac{1}{2\xi\sqrt{(1-\xi^2)}}; \omega_0 = \omega_n\sqrt{(1-2\xi^2)}$$
(7.32)

and also that *resonance* occurs (that is, an amplitude greater than unity appears) only if $\xi < 0.707$. The relationship between the undamped natural frequency ω_n, the natural frequency $\omega_r [= \omega_n\sqrt{(1-\xi^2)}]$, and the resonant frequency ω_0 is shown in figure 7.12; it is seen that, as $\xi \to 0$, the three frequencies all tend to the undamped natural frequency.

Finally, note that the diagram lies in two quadrants whereas that for a first-order lag appeared only in one. Similarly, a third-order system will lie in three quadrants (representing a possible 270° phase lag) and so on as the order of the system increases.

Second-order term in numerator $T^2s^2 + 2T\xi s + 1$. The Nyquist diagram is shown in figure 7.11 (above the real axis) and can be verified as an exercise by the reader.

Example 7.3
Construct the Nyquist diagram for the system with transfer function

$$G(s) = \frac{1 + Ts}{1 + \alpha Ts}$$

Frequency-response Methods

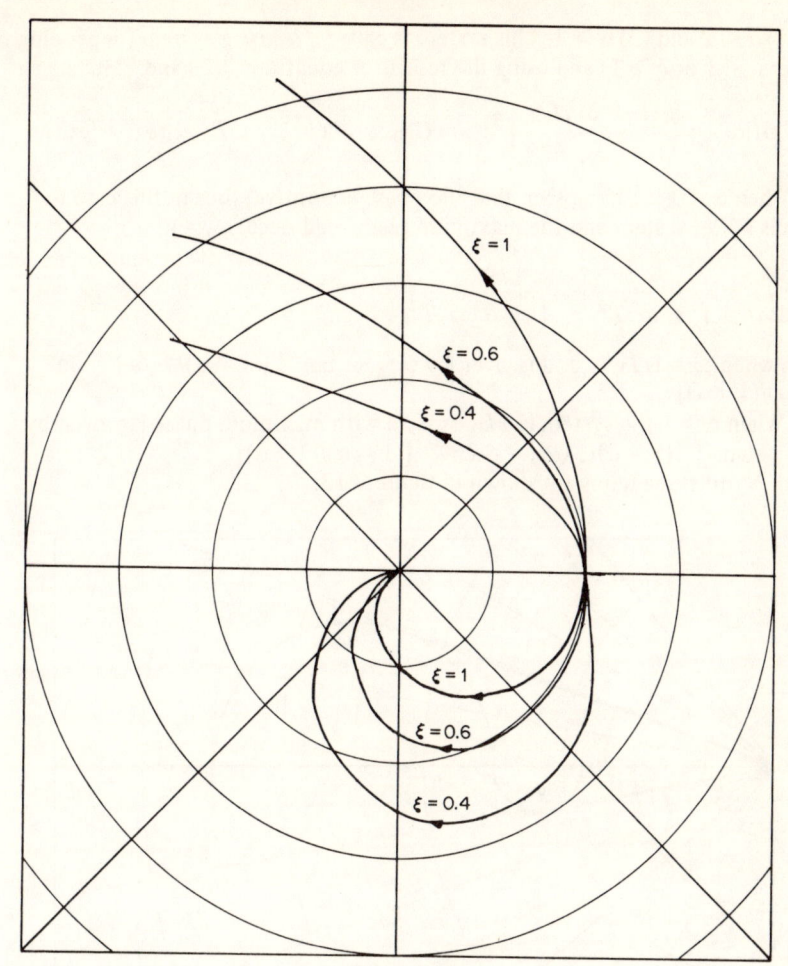

Figure 7.11

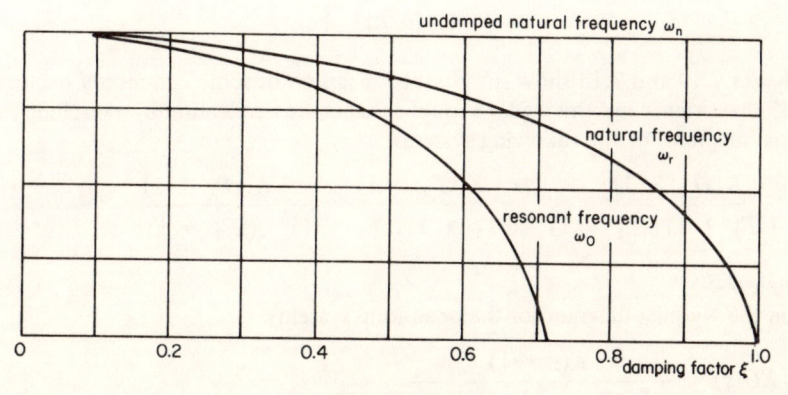

Figure 7.12

when (i) $\alpha < 1$ and (ii) $\alpha > 1$. This system is called a *lead/lag* system (depending on whether $\alpha < 1$ or $\alpha > 1$) and using the results of equations 7.29 and 7.30

$$|G(j\omega)| = \left(\frac{1 + \omega^2 T^2}{1 + \alpha^2 \omega^2 T^2}\right)^{\frac{1}{2}}; \quad \arg G(j\omega) = \tan^{-1} \omega T - \tan^{-1} \alpha \omega T$$

(i) When $\alpha < 1$ it can be seen that the phase is positive; thus in this case the system is a *lead* system and the maximum phase lead occurs when

$$\frac{d\phi}{d\omega} = \frac{T}{1 + \omega^2 T^2} - \frac{\alpha T}{1 + \alpha^2 \omega^2 T^2} = 0$$

that is, when $\omega = 1/T\sqrt{\alpha}$, and is given by $\phi_{max} = \tan^{-1}[(1 - \alpha)/2\sqrt{\alpha}] = \sin^{-1}[(1 - \alpha)/(1 + \alpha)]$.

(ii) When $\alpha > 1$ the system is a *lag* system with maximum phase lag given by $\phi_{max} = -\tan^{-1}[(1 - \alpha)/2\sqrt{\alpha}] = -\sin^{-1}[(1 - \alpha)/(1 + \alpha)]$.

Both Nyquist diagrams are shown in figure 7.13.

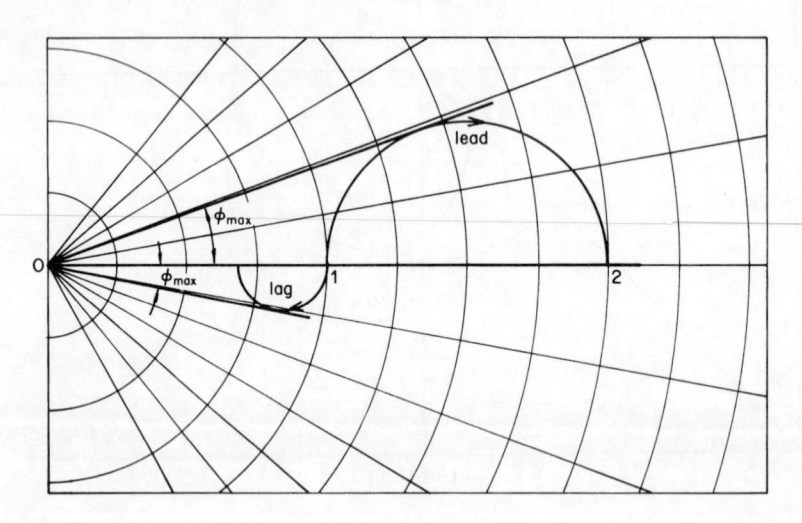

Figure 7.13

Figures 7.14 and 7.15 show the Nyquist diagrams of some commonly occurring open-loop systems and the reader should enhance this collection by sketching the Nyquist diagrams of the following systems

$$\frac{K(sT_3 + 1)}{(sT_2 + 1)(sT_1 + 1)}; \quad \frac{K(sT_3 + 1)}{s(sT_1 + 1)(sT_2 + 1)}; \quad \frac{K(sT_2 + 1)}{s(sT_1 + 1)}$$

Example 7.4
Obtain the Nyquist diagram for the open-loop system

$$KG(s) = \frac{K(s + 1)}{s(s + 2)(s + 3)(s + 4)}$$

Frequency-response Methods

Figure 7.14
(a) $K/(sT_1 + 1)$;
(b) $K/[(sT_1 + 1)(sT_2 + 1)]$;
(c) $K/[(sT_1 + 1)(sT_2 + 1)(sT_3 + 1)]$

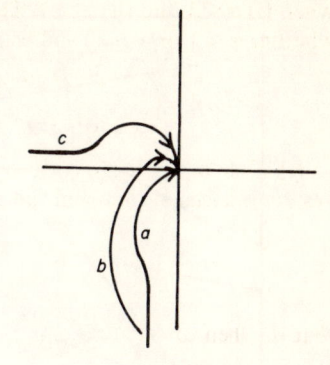

Figure 7.15
(a) $K/[(s(sT_1 + 1)]$;
(b) $K/[s(sT_1 + 1)(sT_2 + 1)]$;
(c) $K/[s^2(sT_1 + 1)]$

when (i) $K = 30$, (ii) $K = 140$, (iii) $K = 200$; comment on the results.
Initially the transfer function is put into the form

$$KG(s) = \frac{K(s + 1)}{(s) 2 (0.5s + 1) 3 (0.33s + 1) 4 (0.25s + 1)}$$

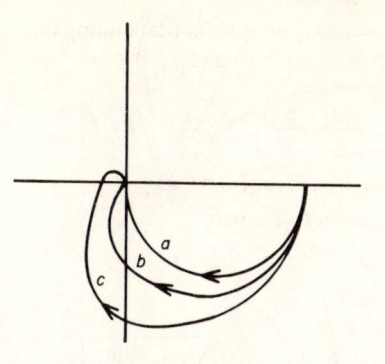

$$= \frac{K^1(s + 1)}{s(0.5s + 1)(0.33s + 1)(0.25s + 1)}$$

where $K^1 = K/24$. From previous results

$$|G(j\omega)| = K^1 \frac{(1 + \omega^2)^{\frac{1}{2}}}{\omega(1 + 0.25\omega^2)^{\frac{1}{2}}(1 + 0.11\omega^2)^{\frac{1}{2}}(1 + 0.063\omega^2)^{\frac{1}{2}}}$$

and

$$\arg G(j\omega) = \tan^{-1} \omega - \tan^{-1} 0.5\omega - \tan^{-1} 0.33\omega - \tan^{-1} 0.25\omega - 90°$$

Therefore, when ω is small, the gain is large and lagging by $90°$; when ω is large, the gain is small and lagging by $270°$. Intermediate points are given in table 7.1 and the corresponding Nyquist diagrams are shown in figure 7.16.

Table 7.1

ω	0	1	2	3	4	5	10	20	∞
$|G(j\omega)|/K^1$	∞	1.16	0.59	0.33	0.2	0.12	0.021	0.0029	0
ϕ	-90	-104	-132	-156	-175	-190	-226	-247	-270

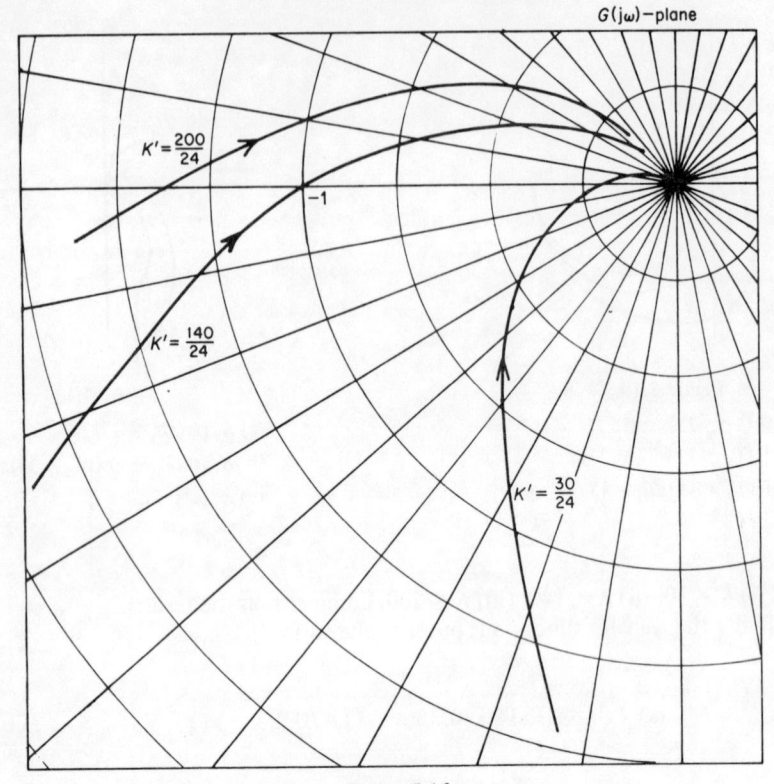

Figure 7.16

When the system is in closed loop it is possible to establish using the Routh criterion that (i) when $K = 30$ the system is stable (see example 6.3) and all the poles are in the left half s-plane; (ii) when $K = 140$ the system is on the threshold of instability, that is, the closed loop has two poles on the imaginary axis; (iii) $K = 200$ the system is unstable and has two poles in the right half s-plane.

It is seen that the Nyquist diagram for $K = 30$ passes to the right of the -1 point in the direction of increasing ω (or conversely the -1 point is on the left), the diagram for $K = 140$ passes through the -1 point and finally that for $K = 200$ passes to the left of the point. It will be shown later that this observation is in fact a criterion for closed-loop stability for a certain class of open-loop systems (known as Nyquist's stability criterion) and can be used to assess the absolute and relative stability of closed-loop systems. Note the importance of the -1 point: it should be evident to the reader why this is!

Non-minimum-phase Systems

The systems considered so far have been stable systems with zeros occurring in the left half of the s-plane. These systems are called *minimum-phase* systems and indicate that *all* denominator terms contribute phase lag whilst numerator terms contribute phase lead.

Frequency-response Methods

Non-minimum-phase systems are those systems that are stable but have zeros in the right half s plane, thus indicating the presence of numerator terms that contribute phase lag to the over-all frequency response. These systems are by their very nature more difficult to control (because of the additional phase lag).

Bode Diagram

The Bode diagram consists of two graphs—of amplitude or gain against frequency and of phase against frequency—and is the non-parametric representation of the Nyquist diagram. Although the Nyquist diagram has the advantage of presenting the complete frequency response compactly in one diagram, the Bode diagram, if logarithmic scales are used for the gain and frequency axes, has the following advantages: (i) the individual gain terms can be added rather than multiplied; (ii) a frequency range of several decades can be considered (one decade means a frequency variation of 10 to 1); (iii) the Bode diagram of complicated systems can be quickly sketched by the use of asymptotic approximations, thus avoiding lengthy calculation. In many cases this sketch will be sufficiently accurate to decide upon initial relative stability.

Thus if

$$G(s) = G_1(s)G_2(s)G_3(s) \tag{7.33}$$

the open-loop frequency response is given by

$$|G(j\omega)| = |G_1(j\omega)| \, |G_2(j\omega)| \, |G_3(j\omega)|$$

and

$$\phi(j\omega) = \phi_1(j\omega) + \phi_2(j\omega) + \phi_3(j\omega) \tag{7.34}$$

If logarithms (to base 10) of the gain equation are taken then

$$\log|G(j\omega)| = \log|G_1(j\omega)| + \log|G_2(j\omega)| + \log|G_3(j\omega)| \tag{7.35}$$

thus confirming one of the advantages listed above. It is usual to express the gain in *decibels* (db) using the relationship

$$\text{gain in decibels (db)} = 20 \log_{10} (\text{actual gain}) \tag{7.36}$$

This practice derives from the telecommunications branch of electrical engineering.

Thus, if the gain is 2, the gain in db is $20 \log_{10} 2$, that is, 6 db; if the gain is 1/2, the gain in db is $20 \log_{10}(1/2)$, that is -6 db. Thus a positive value of the gain in db indicates *magnification* whilst a negative value indicates *attenuation*.

It is now possible to construct Bode diagrams of the standard terms including

$$K, \quad s, \quad as + 1, \quad bs^2 + cs + 1$$

from which the diagrams of more complicated systems may be obtained.

Gain term K. The gain in db is simply $20 \log K$ and is independent of frequency. Thus it corresponds to a line parallel to the frequency axis on the graph of the gain in db against $\log \omega$, as shown in figure 7.17.

Integrating term 1/s. From equation 7.27, $|G(j\omega)| = 1/\omega$ and $\arg G(j\omega) = -90°$, and the gain in db is given by

$$20 \log (1/\omega) = -20 \log \omega \tag{7.37}$$

146   **Introduction to Control Theory**

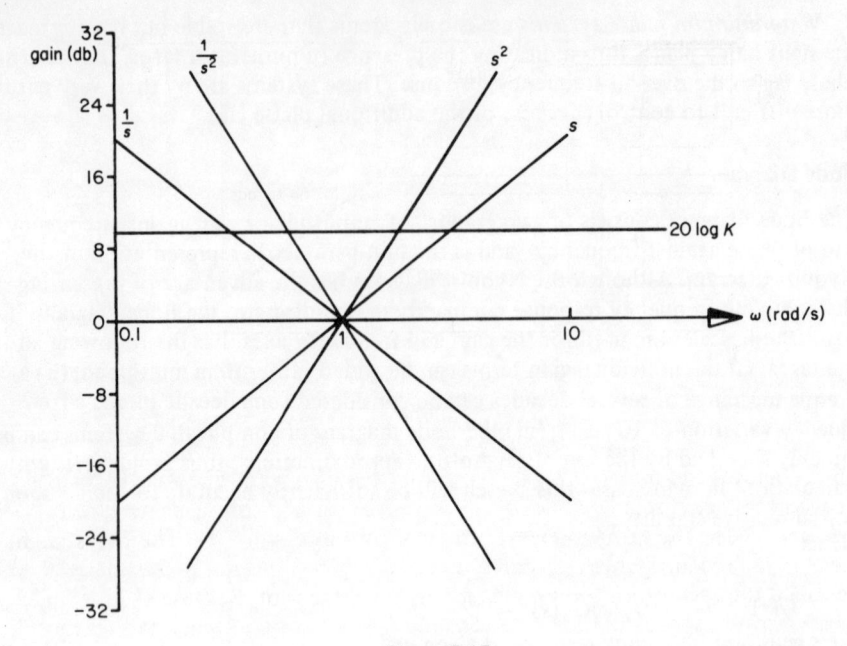

Figure 7.17

This corresponds to a straight line of slope -20 db/decade, (or -6 db/octave) on the graph of gain in decibels against log ω and passes through the $\omega = 1$ point on the frequency axis, as shown in figure 7.17. The graph of phase against log ω is a line parallel to the frequency axis passing through the $-90°$ point on the phase axis. Figure 7.17 also shows the gain curve for $1/s^2$: note the slope of -40 db/decade.

Differentiating term s. From equation 7.28, $|G(j\omega)| = \omega$ and arg $G(j\omega) = +90°$, and the gain in db is $20 \log \omega$, as shown in figure 7.17. The phase-log ω graph is independent of frequency and passes through the $+90°$ point on the phase axis.

First-order lag term $1/(Ts + 1)$. From equation 7.29 the gain in decibels is given by

$$20 \log \left(\frac{1}{1 + \omega^2 T^2}\right)^{\frac{1}{2}} = -20 \log (1 + \omega^2 T^2)^{\frac{1}{2}}$$

$$= -10 \log (1 + \omega^2 T^2) \tag{7.38}$$

and the phase angle by

$$\arg G(j\omega) = -\tan^{-1} \omega T \tag{7.39}$$

When ω is small, $\omega^2 T^2$ will be negligible compared with unity and hence the gain in db will be a close approximation to $-10 \log_{10} 1$, that is, 0 db. Thus, at the lower end of the frequency spectrum the gain in db will be zero. For higher frequencies, the $\omega^2 T^2$ term will dominate and the gain will be $-10 \log \omega^2 T^2$, that is, $-20 \log \omega - 20 \log T$, which corresponds to a straight line of slope -20 db/decade (or

Frequency-response Methods 147

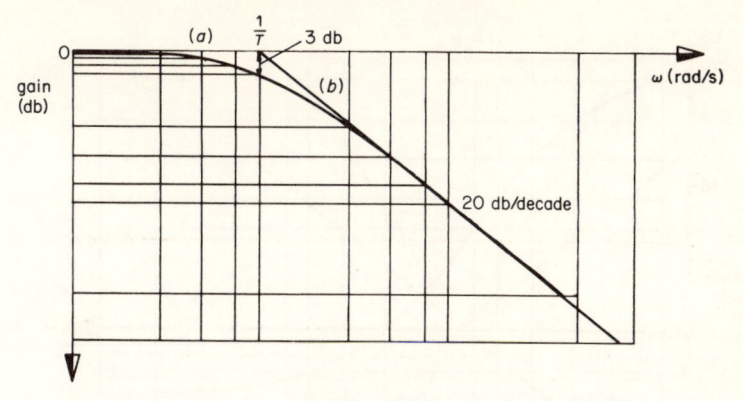

Figure 7.18

−6 db/octave) and intersect $\omega = 1/T$ on the frequency axis. The frequency $\omega = 1/T$ is called the *break frequency* or *corner frequency*, and at this value the gain is −10 log 2 or −3 db. The gain–frequency plot is shown in figure 7.18. The two straight lines (a) and (b) in figure 7.18 represent a good approximation to the true plot and are called the *asymptotic approximation* to the Bode plot. Because of its simplicity the asymptotic approximation is used by designers to obtain quick and (in most cases) adequate representations of complex systems. The difference in gain between the two plots is highlighted in **Table 7.2**

ωT	0.1	0.3	0.5	0.75	1	3	5	7.5	10	50
True gain	−0.043	−0.37	−0.97	−1.94	−3	−10	−14.1	−17.6	−20	−34
Approximate gain	0	0	0	0	0	−9.5	−14.0	−17.5	−20	−34
Phase	−5.7	−16.7	−26.6	−36.9	−45	−71.6	−78.7	−82.4	−84.3	−88.9

The plot of phase angle against log ω is shown in figure 7.19 and, although it is not possible to use straight-line approximations, the curve is symmetrical about the $\omega T = 1$, $\phi = -45°$ point. The plot confirms that a first-order lag experiences only small phase lag at low frequencies, rising to 90° lag at high frequencies.

First-order lead term, $Ts + 1$. The complete Bode diagram with the asymptotic approximation is shown in figure 7.20.

Second-order term in denominator $1/(T^2 s^2 + 2T\xi s + 1)$. From equation 7.31 the gain is db is given by

$$20 \log \left[\frac{1}{(1 - u^2)^2 + (2\xi u)^2} \right]^{\frac{1}{2}} = -10 \log [(1 - u^2)^2 + (2\xi u)^2] \quad (7.40)$$

and the phase angle by

$$\arg G(j\omega) = \tan^{-1} \left[-\frac{2\xi u}{1 - u^2} \right] \quad (7.41)$$

where $T = 1/\omega_n$, $u = \omega/\omega_n$, that is, $u = \omega T$.

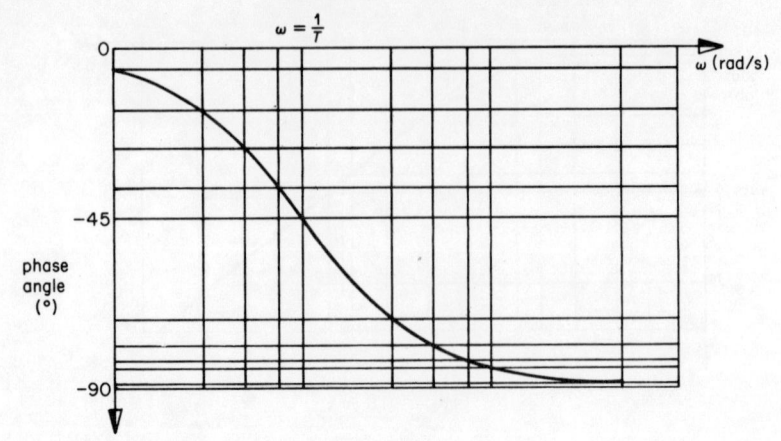

Figure 7.19

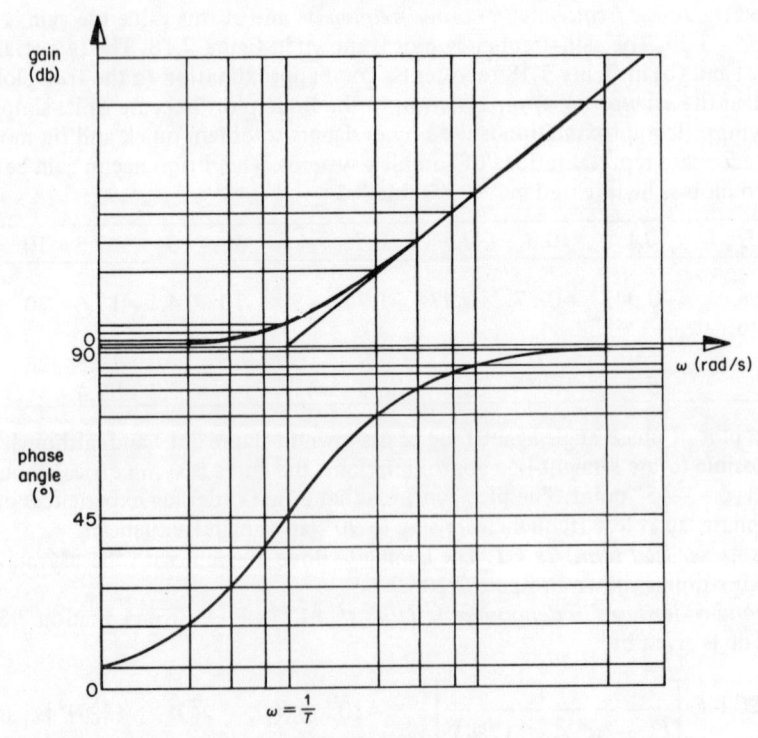

Figure 7.20

At low frequency, the gain is $-10 \log 1$, that is, 0 db, whilst at high frequencies, the gain is $-10 \log u^4$, that is, $-40 \log \omega - 40 \log T$, which corresponds to a straight line of slope -40 db/decade (this is correct since it is a second-order system) intersecting the frequency axis at $\omega T = 1$, that is, $\omega = \omega_n$. These lines represent the asymptotic approximation to a second-order system. At the break frequency $\omega = \omega_n$, the actual gain in db is $-10 \log 4\xi^2$, that is, $-20 \log 2\xi$, or $20 \log (1/2\xi)$. Thus, when $\xi = 0.5$, the gain is 0 db, and hence the gain passes through the break-frequency point at this value of damping. For low damping ($\xi < 0.5$) it is seen that the asymptotic approximation can be in appreciable error at and around the break frequency. This is highlighted in figure 7.21a, which shows the gain–frequency plot: note also that resonance only occurs if $\xi < 0.707$ and the resonance frequency ω_0 is given by $\omega_0 T = \sqrt{(1 - 2\xi^2)}$ as pointed out earlier.

The plot of phase angle against frequency is shown in figure 7.21b: at low frequencies only small phase lags are present, whilst at high frequencies the output lags the input by 180°. At the break frequency the phase angle is −90°.

Second-order term in numerator $T^2s^2 + 2T\xi s + 1$. The Bode diagram follows readily from the Bode diagram of figure 7.21 and is left as an exercise for the reader.

(a)

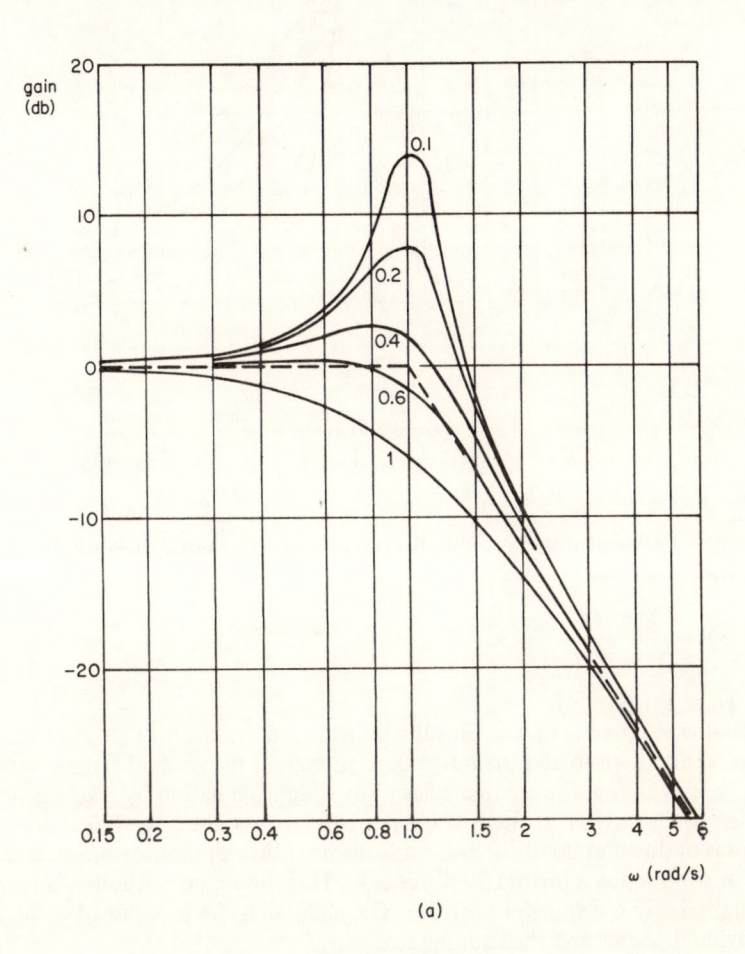

Figure 7.21

(b)

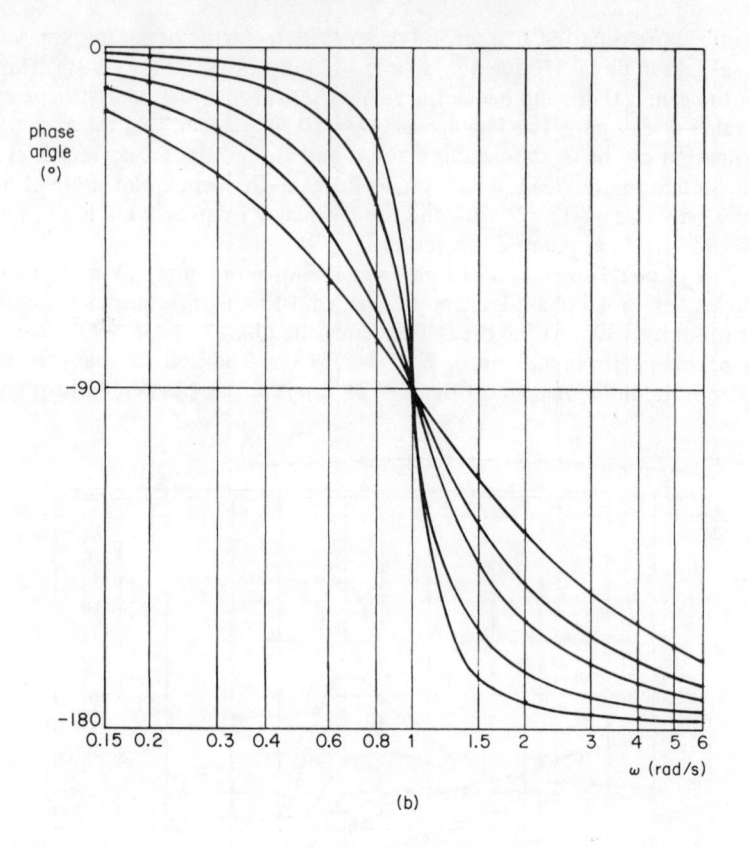

Figure 7.21 (cont.)

Example 7.5
Making use of asymptotic approximations, construct the Bode diagram of the system with transfer function

$$G(s) = \frac{1 + Ts}{1 + \alpha Ts}$$

when (i) $\alpha < 1$; (ii) $\alpha > 1$.

All that is required is first to plot the individual diagrams for $(1 + Ts)$ and $1/(1 + \alpha Ts)$ and then to add them together. In fact, as the reader becomes more skilled, he may obtain the complete asymptotic approximation by starting at zero gain, then having arrived at the first break point (that is, the lowest frequency) moving up or down at 20 db/decade until the next break point is reached and then adding or subtracting a further 20 db/decade. This should be continued for each individual term in the transfer function. The phase diagram is obtained by drawing the individual phases and then adding together.

(i) When $\alpha < 1$ the first break point is with respect to the numerator at $\omega = 1/T$; the asymptotic approximation rises at 20 db/decade until the next break point at

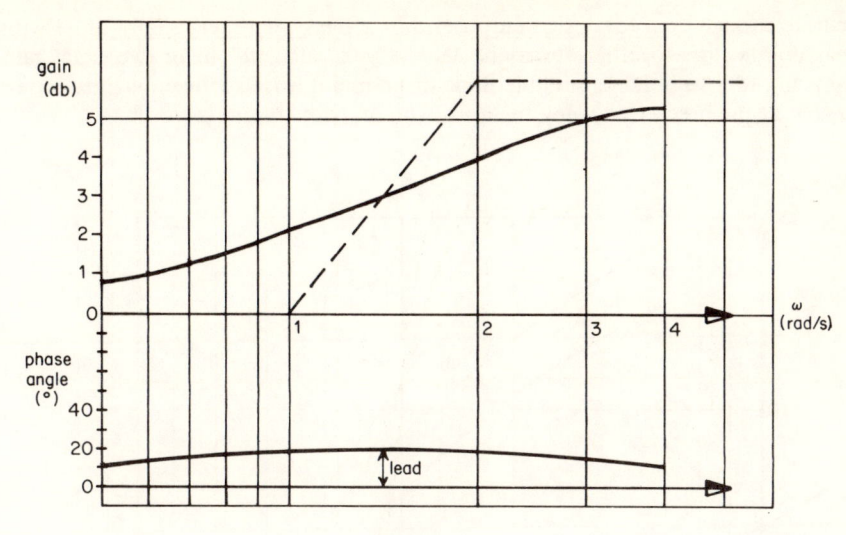

Figure 7.22

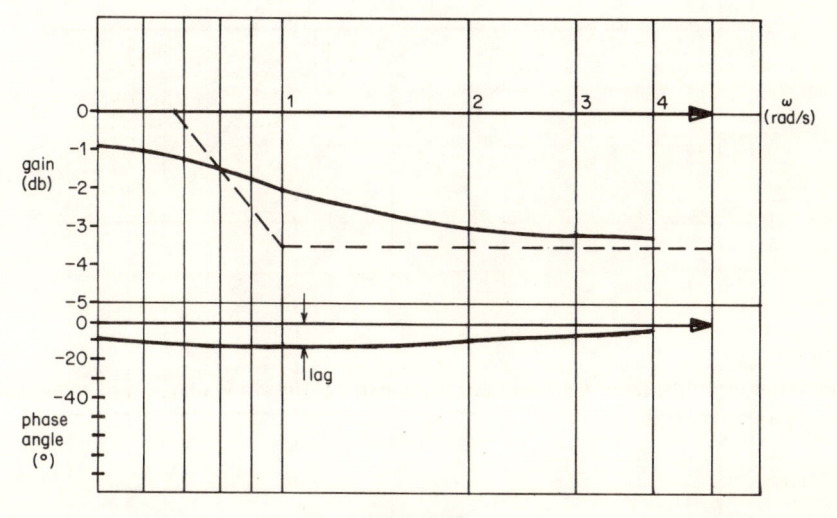

Figure 7.23

$\omega = 1/\alpha T$ when 20 db/decade is subtracted due to the denominator and the plot becomes parallel to the frequency axis. The resulting Bode diagram is shown in figure 7.22 and clearly indicates the *leading* nature of the system.

(ii) When $\alpha > 1$, the denominator reaches its break frequency first at $\omega = 1/\alpha T$, thus causing the asymptote to drop at 20 db/decade until the break frequency at $\omega = 1/T$, after which the gain becomes parallel to the frequency axis. The Bode diagram is in figure 7.23.

In example 7.5 the asymptotic approximations are seen to give a fairly adequate representation of the actual Bode diagrams and in many cases the approach is

152 Introduction to Control Theory

accurate enough even for design purposes. For some systems, especially those with break points close together, it may be necessary to calculate one or two exact gain points in order to obtain a suitable Bode diagram but it is usually enough to remember that at the break frequency the gain of an individual term is ± 3 db.

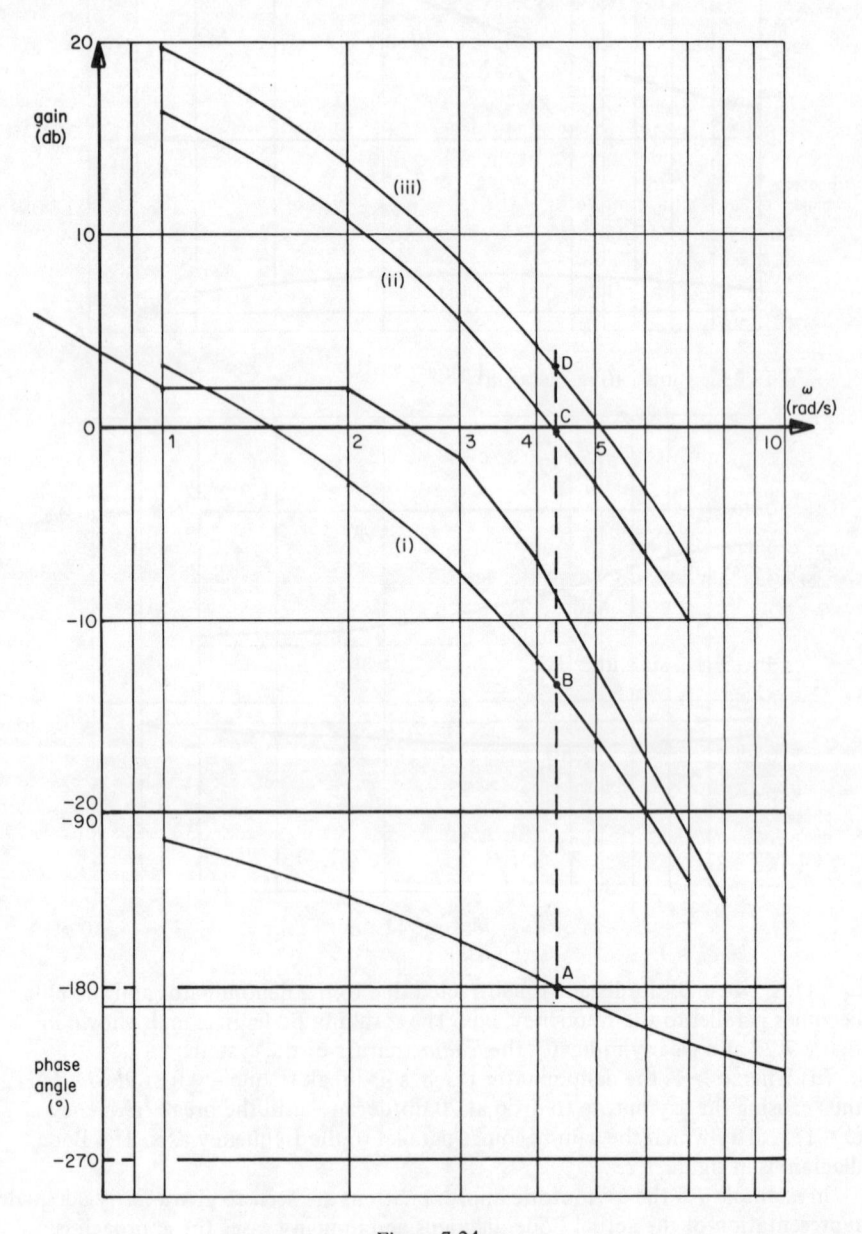

Figure 7.24

Example 7.6
Obtain the Bode diagram for the open-loop system

$$KG(s) = \frac{K(s + 1)}{s(0.5s + 1)(0.33s + 1)(0.25s + 1)}$$

when (i) $K = 1.25$; (ii) $K = 5.83$; (iii) $K = 8.33$.

The three Bode diagrams are shown in figure 7.24, where it is seen that the actual gain plot follows the general asymptotic shape. Referring back to example 7.4, it was stated that case (i) represents a stable closed-loop system, (ii) a system on tne tnreshold of instability and (iii) an unstable closed-loop system. This result can be interpreted from the Bode diagram in the following way.

Case (i): at a phase angle of $-180°$ (point A), where $\omega = 4.35$ rad/s, the gain is -14 db (point B), that is, 0.2. By previous reasoning, since this is less than unity, the system is stable.
Case (ii): the gain is 0 db, that is, 1.0, and the closed-loop system will oscillate continuously.
Case (iii): the gain is 3 db, that is, 1.4, and the signal around the loop will continue to grow, indicating instability.

More will be said later about the use of the Bode diagram to assess the absolute and relative stability of closed-loop systems.

Nichols Diagram

An alternative parametric representation of the frequency response of a system is the Nichols diagram, on which the gain in decibels is plotted against phase using rectangular coordinates with ω as a parameter. Nichols diagrams are mainly used for predicting closed-loop-system performance from open-loop-system behaviour and hence relative stability. The Nichols diagram for a second-order lag is shown in figure 7.25.

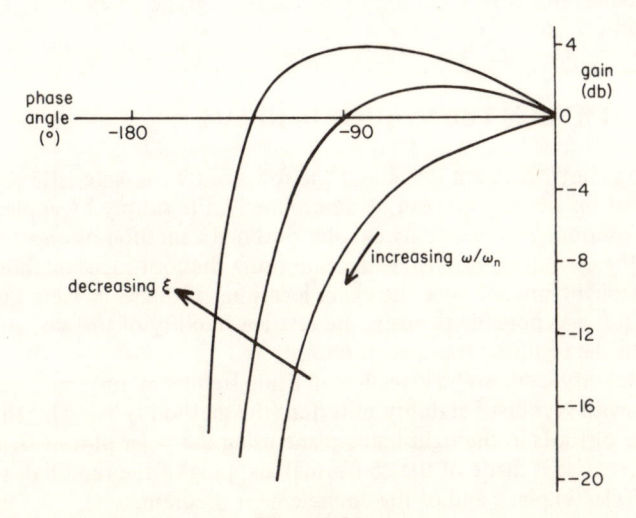

Figure 7.25

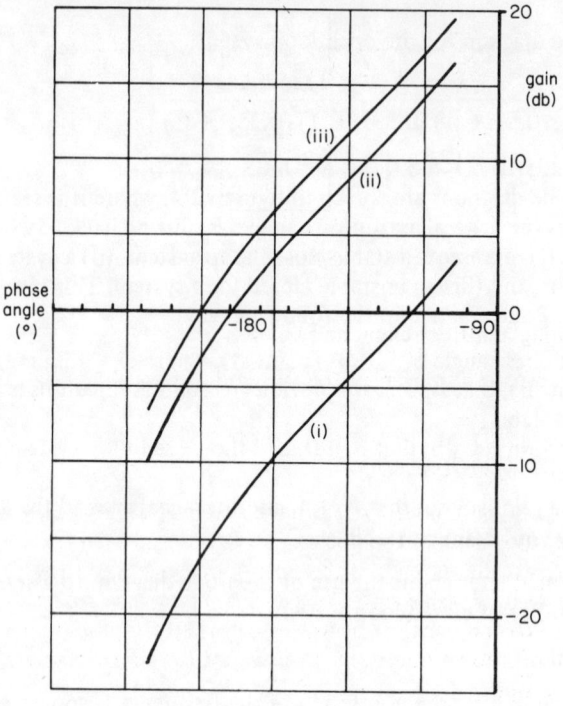

Figure 7.26

Example 7.7
The Nichols diagrams corresponding to the Bode diagrams of example 7.6 are shown in figure 7.26.

7.3 STABILITY FROM THE NYQUIST DIAGRAM

A closed-loop control system is stable if the roots of its characteristic equation lie in the left half s-plane and this can be determined quite simply by application of the Routh criterion. The preceding chapter outlined a method of assessing the stability of the closed-loop transfer function using the root-locus method, which was in fact straightforward once the exact locations of the roots were known. Furthermore, it was possible to assess the relative stability of the closed-loop system and to obtain the resulting transient response.

System stability can also be investigated using frequency-response methods in conjunction with Nyquist's stability criterion; the method is basically to determine the existence of roots in the right half s-plane using the polar plot of $G(j\omega)H(j\omega)$ for $0 \leqslant \omega < \infty$. Use is made of the conformal mapping of the right half s-plane on to the $1 + G(s)H(s)$ plane and of the encirclement theorem.

Frequency-response Methods   155

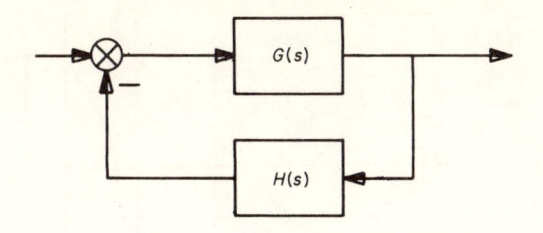

Figure 7.27

Conformal Mapping and Encirclement Theorem

The closed-loop transfer function of the system in figure 7.27 is

$$G'(s) = \frac{G(s)}{1 + G(s)H(s)} \qquad (7.42)$$

in which the zeros of $1 + G(s)H(s)$ determine system stability. The poles of $1 + G(s)H(s)$ are equivalent to the poles of $G(s)H(s)$ and even if $G(s)$ has poles in the right half s-plane, that is, even if the open-loop system is unstable, the closed-loop system may still be stable since $1 + G(s)H(s)$ may have no zeros in the right half s-plane. Since $G(s)$ is usually known in factored form, the numbers of unstable poles in $G(s)$ will be known but, even if this is not the case, they can be determined using the Routh criterion.

Consider now the pole-zero configuration of $1 + G(s)H(s)$, where

$$1 + G(s)H(s) = \frac{K(s - z_1)(s - z_2) \ldots}{s^l(s - p_1)(s - p_2) \ldots} \qquad (7.43)$$

which is shown in figure 7.28. Choose a point T in the right half s-plane and allow it to move in a clockwise direction over a closed contour C that neither encloses any poles or zeros nor passes through any of them (figure 7.29a). This will result in vectors such as $(s - p_4)$ changing direction but giving *no net rotation* of $1 + G(s)H(s)$. If the exercise is repeated but this time the contour C is allowed to enclose two zeros and one pole (figure 7.29b), the *net* change in angle of the vectors $(s - z_1)$,

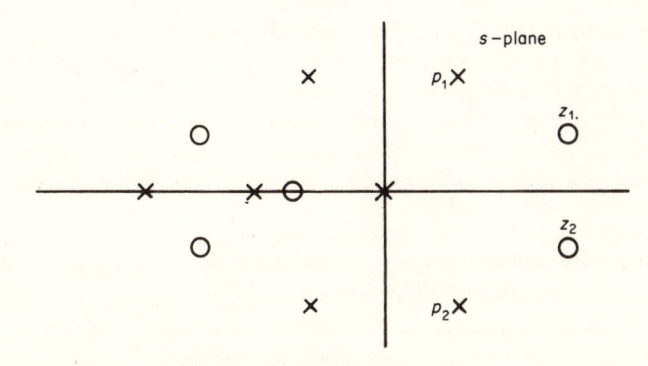

Figure 7.28

Introduction to Control Theory

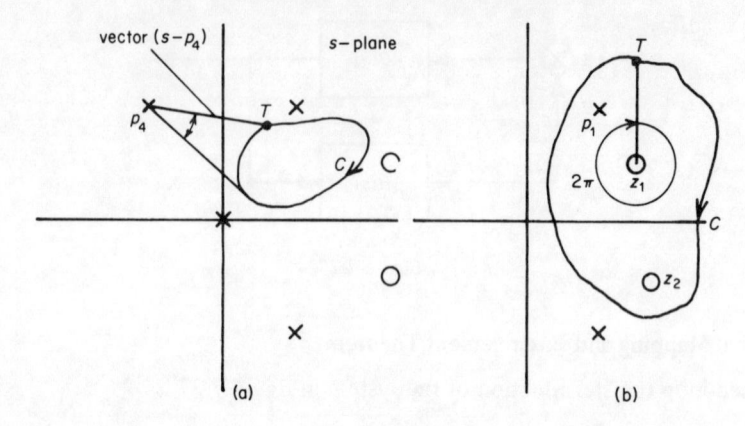

Figure 7.29

$(s - z_2)$ and $(s - p_1)$ is 2π rad clockwise, resulting in a net rotation of $1 + G(s)H(s)$ by $(2 \times 2\pi - 1 \times 2\pi)$, that is, 2π rad, in a clockwise direction.

Thus in general if the contour C encloses Z zeros and P poles then the net rotation of $1 + G(s)H(s)$ is $(Z - P)\, 2\pi$ rad clockwise.

As the point T traces a clockwise closed contour C in the s-plane, the function $1 + G(s)H(s)$ will also trace a closed contour in a direction dependent on $1 + G(s)H(s)$, if T encloses no poles or zeros or an equal number of them, $1 + G(s)H(s)$ will not encircle the origin, whilst if T encloses a pole then $1 + G(s)H(s)$ will encircle the origin once, counter-clockwise (see figure 7.30).

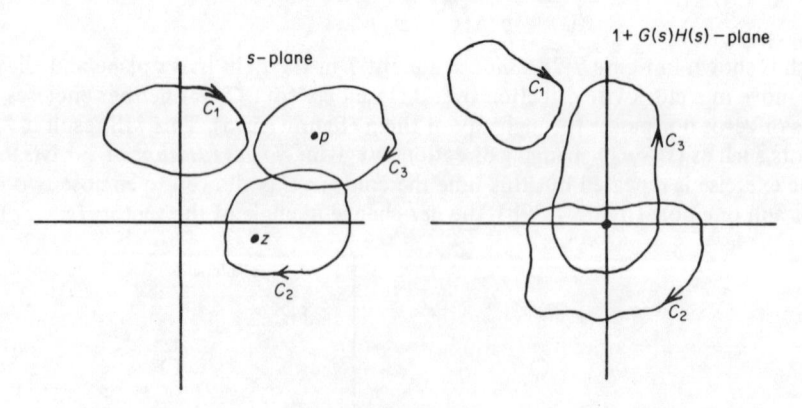

Figure 7.30

Therefore, if the contour encloses Z zeros and P poles in the s-plane, the net rotation of $1 + G(s)H(s)$ about the origin is

$$\left. \begin{array}{l} 2\pi(Z - P) \text{ rad clockwise, or} \\ 2\pi(P - Z) \text{ rad counter-clockwise or positive} \end{array} \right\} \qquad (7.43)$$

Frequency-response Methods 157

Nyquist Stability

Allow s to trace on the s-plane a contour C which encloses all the zeros Z and all the poles P of $1 + G(s)H(s)$ in the finite part of the right half s-plane but excludes the origin; that is, C is arranged to enclose *all* the right half s-plane by letting $R_1 \to 0$ and $R_2 \to \infty$, as shown in figure 7.31. Note that for physical systems $G(s)H(s) \to 0$

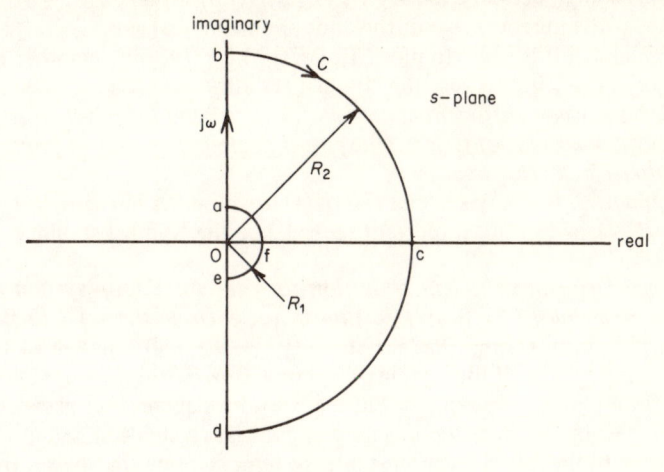

Figure 7.31

as $s \to \infty$. Hence the large semicircle R_2 corresponds to the point $1 + j0$ on the $1 + G(s)H(s)$ plane, and, as $R_1 \to 0$, $G(s)H(s)$ tends to a finite and *real* point or to infinity. Thus, the path followed by s may be described by the following paths

ab: $\quad s = j\omega \quad (0 \leqslant \omega < \infty)$

bcd: $\quad s = \lim_{R \to \infty} Re^{j\theta} \quad \dfrac{\pi}{2} \leqslant \theta \leqslant -\dfrac{\pi}{2}$

de: $\quad s = j\omega \quad (-\infty < \omega \leqslant 0)$

efa: $\quad s = \lim_{R \to 0} Re^{j\theta} \quad -\dfrac{\pi}{2} \leqslant \theta \leqslant \dfrac{\pi}{2}$

Note that the paths ab and de correspond to the Nyquist diagram of $1 + G(j\omega)H(j\omega)$ $(-\infty < \omega < \infty)$ and the complete path corresponds to the *full* Nyquist diagram.

It is now possible using the mapping of $1 + G(j\omega)H(j\omega)$ and the encirclement theorem, to determine system stability in the following way.

Nyquist stability criterion. If s describes the contour C defined by figure 7.31, that is, a contour enclosing all the poles P and zeros Z of $1 + G(s)H(s)$ that have positive real parts, then the number of times the locus (or polar plot) of $1 + G(j\omega)H(j\omega)$ encircles the origin in a *clockwise* sense in a *stable* system is

$$N = -P \tag{7.44}$$

In other words, the polar plot of $1 + G(j\omega)H(j\omega)$ must encircle the origin P times in a counter-clockwise direction if $1 + G(s)H(s)$ is to have *no* zeros in the right half s-

plane and hence if the closed-loop transfer function $G(s)/[1 + G(s)H(s)]$ is to have *no* unstable poles.

Thus to study stability it is only necessary to obtain the polar plot of $1 + G(j\omega)H(j\omega)$.

Comment 1. Since $G(j\omega)H(j\omega) = G(-j\omega)H(-j\omega)$ for a system with real coefficients, it is only necessary to plot $1 + G(j\omega)H(j\omega)$ for $0 \leq \omega < \infty$; the remainder of the locus is the mirror image of this plot about the real axis.

Comment 2. It is usual to plot $G(j\omega)H(j\omega)$ for $0 \leq \omega < \infty$ rather than $1 + G(j\omega)H(j\omega)$. In this case the Nyquist stability criterion may be stated as follows. *A necessary and sufficient condition for the closed-loop system to be stable is that the polar plot of $G(j\omega)H(j\omega)$ must encircle the $-1 + j0$ point P times in a counter-clockwise direction.*

Comment 3. In many systems $G(s)H(s)$ has no poles in the right half s-plane, that is, $P = 0$, and in this important special case the Nyquist stability criterion is as follows.

A necessary and sufficient condition for the closed-loop system to be stable is that the polar plot of $G(j\omega)H(j\omega)$ must not encircle the $-1 + j0$ point. This can be interpreted by stating that for stability the $-1 + j0$ point must lie to the *left* of the *nearest point* of the Nyquist diagram $G(j\omega)H(j\omega)$ ($0 \leq \omega < \infty$) when moving in the direction of increasing ω. Thus, a Nyquist diagram that passes to the right of the -1 point in the direction of increasing ω gives a stable closed-loop system; one that passes to the left gives an unstable system; and one that passes through the -1 point indicates a closed-loop system on the threshold of instability. The reader should refer to example 7.4.

Conditional Stability

It is possible for a closed-loop system to be stable for a particular value of gain K, but to become unstable as K is either increased or decreased. Such a system is said to be *conditionally stable*; a typical Nyquist diagram and the corresponding root-locus diagram are shown in figure 7.32. Thus, when applying Nyquist's stability criterion (see comment 3) it is essential to consider a point on the curve nearest to the $-1 + j0$ point.

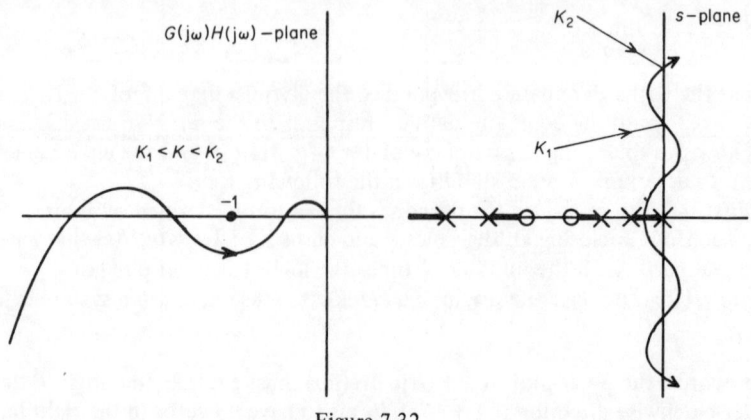

Figure 7.32

Example 7.8
Investigate the stability of the system with forward transference $K/(s-1)$ using frequency-response methods. Confirm the results using the root-locus method.

In this case

$$G(s) = \frac{K}{s-1}$$

Therefore

$$G(j\omega) = \frac{K}{j\omega - 1}$$

and

$$|G(j\omega)| = \frac{K}{\sqrt{(\omega^2 + 1)}}; \quad \arg G(j\omega) = \tan^{-1}\left(\frac{-\omega}{-1}\right) = \pi + \tan^{-1}\omega$$

When $\omega = 0$, it is seen immediately that the gain is K and the phase angle $180°$, and so to investigate stability it is necessary to consider the Nyquist diagrams for $K < 1$, $K = 1$ and $K > 1$. These are shown in figure 7.33a, together with the mirror-images for frequencies in the range $-\infty < \omega \leq 0$. Since the open-loop transfer function has one pole in the right half s-plane it follows from equation 7.44 (using comment 2) that for stability the Nyquist diagram must encircle the $-1 + j0$ point *once* in a counter-clockwise direction. Thus, when $K = 0.5$ (that is, $K < 1$) it is seen that the full Nyquist diagram does not encircle the -1 point and hence the system is unstable. When $K = 1.5$ (that is, $K > 1$) the full Nyquist diagram does encircle the -1 point once in a counter-clockwise direction and hence the system is stable. These results are confirmed by the root-locus diagram of figure 7.33b.

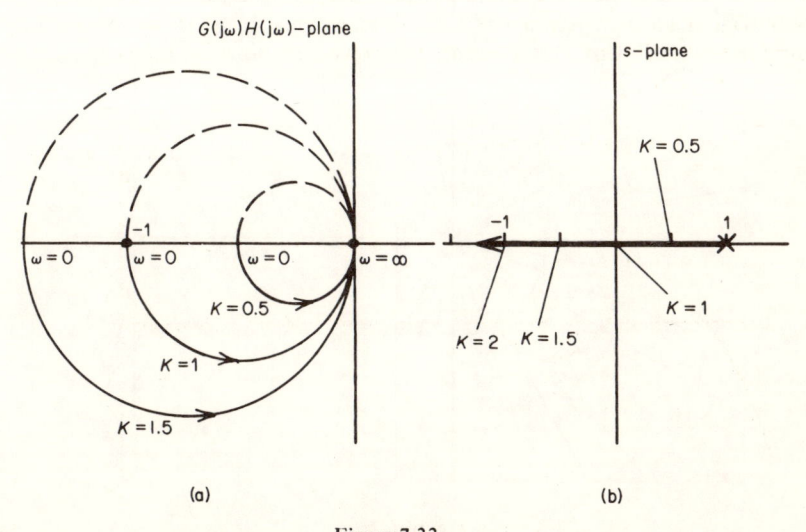

Figure 7.33

Example 7.9

Show that the unity-feedback system with forward transference $K/s^2(s+1)$ is inherently unstable using frequency-response methods. Interpret the results on a Bode diagram and sketch the root-locus diagram.

In this case

$$G(s) = \frac{K}{s^2(s+1)}$$

Therefore

$$G(j\omega) = \frac{K}{-\omega^2(j\omega+1)}$$

and

$$|G(j\omega)| = \frac{K}{\omega^2\sqrt{(1+\omega^2)}}; \quad \arg G(j\omega) = -\pi - \tan^{-1}\omega$$

The Nyquist diagram for $|G(j\omega)|/K$ is shown in figure 7.34a and the root-locus diagram in figure 7.34b. It is seen that the Nyquist diagram passes to the left of the $-1+j0$ point when moving in the direction of increasing ω and thus the system is unstable; since the gain K only affects the magnitude it is unstable for all values of K. This is confirmed by the root-locus diagram of figure 7.34b.

The Bode diagram is shown in figure 7.35. To interpret Nyquist stability using this diagram it is necessary to consider the gain corresponding to a phase lag of 180°. If the gain is greater than unity, that is, 0 db, the system is unstable; if the gain is less than unity, the system is stable. Thus, in the example when the phase lag is 180°, the gain is obviously greater than unity, confirming that the closed-loop system is unstable.

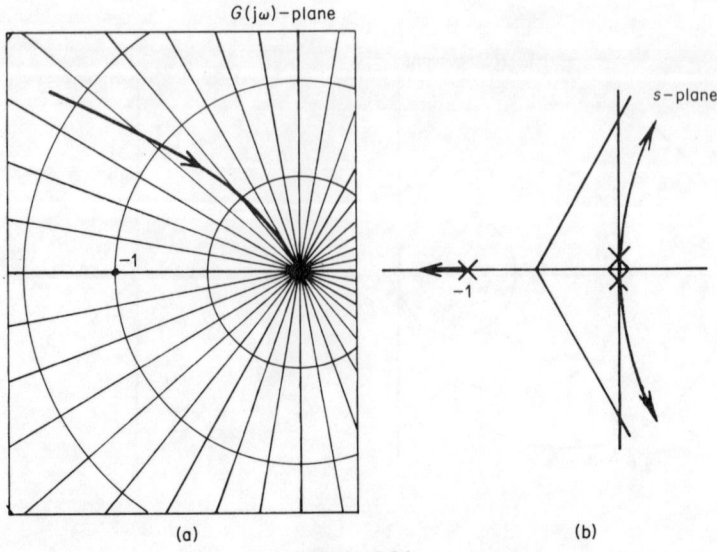

Figure 7.34

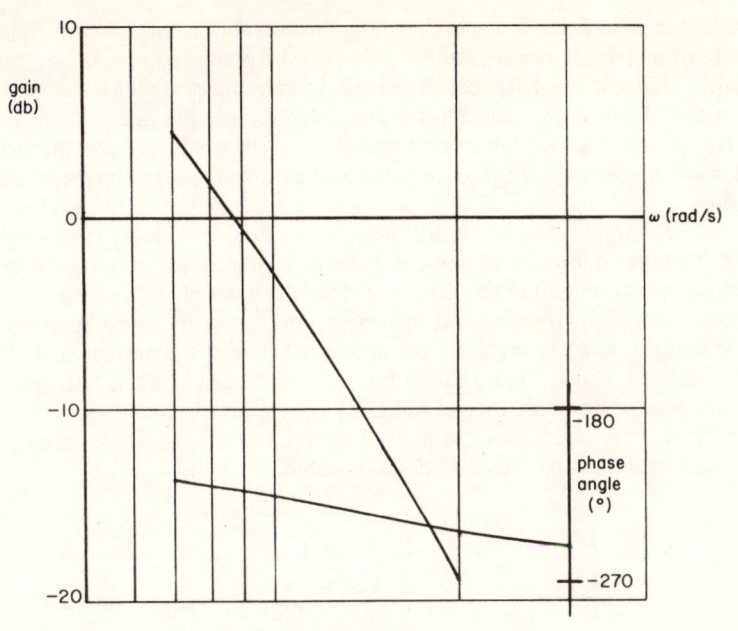

Figure 7.35

7.4 RELATIVE STABILITY: GAIN AND PHASE MARGINS

It is known that zero relative stability occurs when $G(j\omega)H(j\omega) = -1$, that is, when $G(j\omega)H(j\omega) = 1 \angle -180°$, or in other words when the Nyquist diagram passes through the $-1 + j0$ point. Thus, the distance from the critical point to the Nyquist diagram is a measure of the relative stability of the closed-loop system; in general, if this distance is increased, then so is the stability. The degree of relative stability may be expressed quantitatively in terms of the gain and phase of the system by two quantities known as the *gain margin* and *phase margin*.

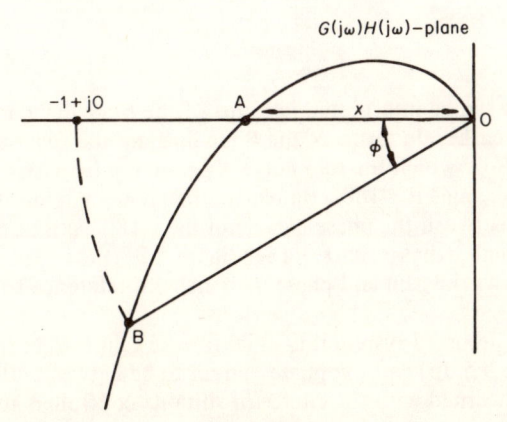

Figure 7.36

162   Introduction to Control Theory

The gain margin is the amount by which the actual gain must be multiplied before the onset of instability occurs assuming that all the phase vectors remain fixed. Thus, in figure 7.36 if x is the distance OA, where A is the point at which the Nyquist diagram cuts the negative real axis (or the $-180°$ vector) the gain margin is defined to be $1/x$, or $-20 \log x$ db, since multiplication of the actual gain by this amount would cause the Nyquist diagram to pass through the $-1 + j0$ point, resulting in instability.

The phase margin is the amount of negative phase shift (that is, phase lag) that must be introduced into the system, without gain change, to cause the Nyquist diagram to pass through the critical point. The phase margin ϕ is obtained by drawing a unit circle with centre at the origin, and taking the angle between the negative real axis and the vector to the intersect between the unit circle and the Nyquist diagram, that is, vector **OB** in figure 7.36; its sign is taken to be positive. Obviously, a negative phase margin indicates system instability.

Some care must be exercised in the calculation of gain and phase margins when dealing with systems that are conditionally stable.

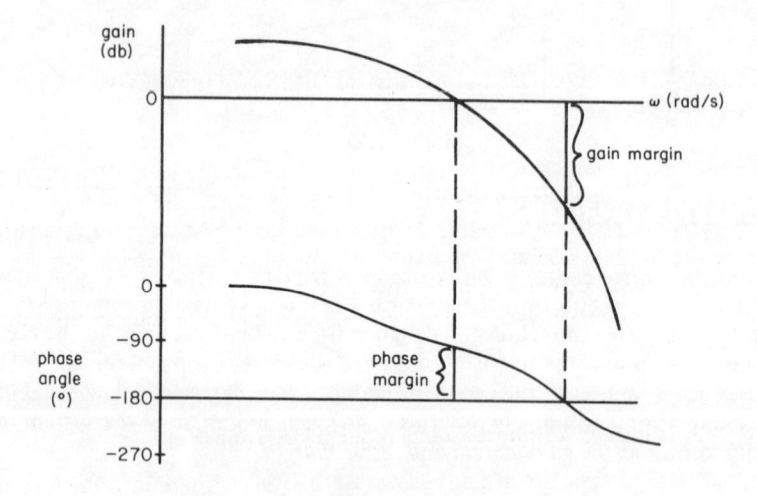

Figure 7.37

Both quantities can be interpreted in terms of the Bode diagram and Nichols diagram discussed earlier. In terms of the Bode diagram the gain margin is the *number of decibels* by which the gain curve is *below* the 0-db axis at the frequency for which the phase angle is $-180°$. Similarly, the phase margin is the number of degrees (positive) between the phase curve and the $-180°$ axis at the frequency for which the gain is 0 db, (that is, unity). (See figure 7.37.)

Figure 7.38 shows the gain and phase margins with reference to the Nichols diagram.

For a correctly designed system it is usual for the gain margin to be between 2 and 3 (that is, 6 to 9.5 db) and the phase margin to be between 40° and 60°. The next chapter is concerned with the choice of suitable controllers to give desired gain and phase margins.

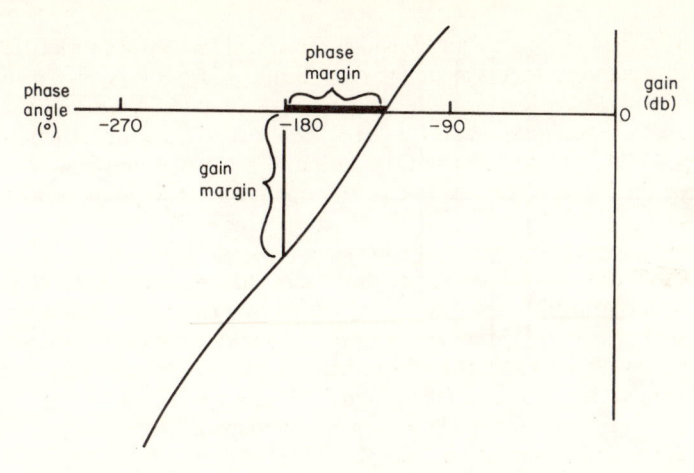

Figure 7.38

7.5 CLOSED-LOOP PERFORMANCE FROM OPEN-LOOP FREQUENCY CHARACTERISTICS

With the root-locus method it is possible to obtain the closed-loop performance in the form of the transient response of the closed-loop system. Similarly, by superimposing curves of constant *closed-loop* amplitude and phase on either the Nyquist diagram or the Nichols diagram, a measure of closed-loop performance can be obtained from the frequency response. Also, it is possible by using a simple construction to obtain an estimate of the poles that dominate the transient response of the closed-loop system.

Approximate Transient Response from Frequency Response

When $s = j\omega$, $0 \leq \omega < \infty$, the imaginary axis of the s-plane maps into the Nyquist diagram of the $G(s)H(s)$ plane as shown in figure 7.39a and b. Now when $s = \sigma + j\omega$, $0 \leq \omega < \infty$, the conformal mapping produces the dotted line labelled $\sigma = -\sigma_1$ in figure 7.39b, and similarly when $s = -\sigma_2 + j\omega$, $-\sigma_3 + j\omega$, etc. This correspondence between the two planes can be used together with a construction due to Campbell to determine the poles that dominate the closed-loop transient response, as follows.

Draw the Nyquist diagram as shown in figure 7.40 and mark on values of frequency at constant intervals in the vicinity of the -1 point and then, using these intervals as base lines, construct a series of equilateral triangles and join their apices. The heights of the triangles will be approximately $0.866\delta\omega$. Using this line as a new base line, the method is repeated until the -1 point is enclosed by an equilateral triangle. From this critical point a line is drawn orthogonal to these base lines to meet the Nyquist diagram at a frequency ω. Also, the number of base lines between the critical point and the Nyquist diagram gives σ.

Thus the poles that dominate the transient response are $-\sigma \pm j\omega$.

164   Introduction to Control Theory

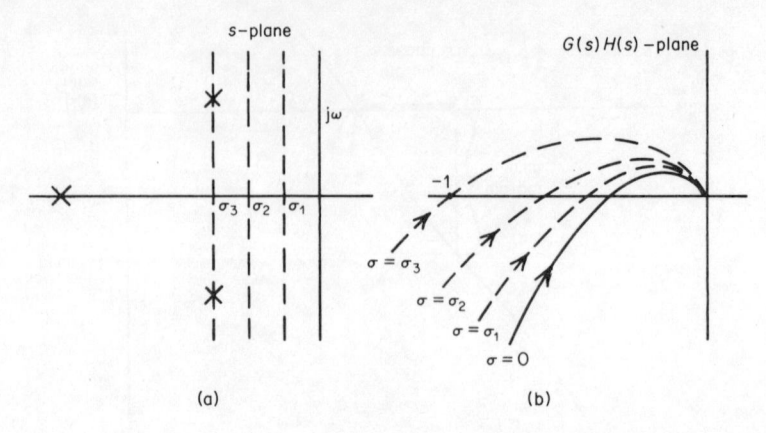

Figure 7.39

Example 7.10
The forward transference of a unity-feedback system $KG(s)$ is given by

$$KG(s) = \frac{10}{(s+3)(s^2+2s+2)}$$

Draw the Nyquist diagram and use it to obtain an estimate of the closed-loop poles. Also obtain the gain margin.

The Nyquist diagram is shown in figure 7.40 together with the construction of equilateral triangles using frequency intervals of 0.2 rad/s. From the figure, $\sigma = 0.2 \times 0.866 \times 3.3 = 0.57$ and $\omega = 2.07$. Thus, the dominant poles are $-0.57 \pm j2.07$ and this is confirmed on the root-locus diagram shown in figure 6.13 (for $K = 10!$).

Also from figure 7.41, $x = 0.29$; therefore the gain margin is 3.45, indicating that the gain could be increased from 10 to 34.5 before the threshold is reached. This is also confirmed in figure 6.13.

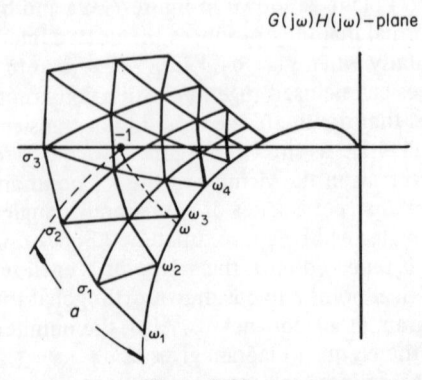

Figure 7.40

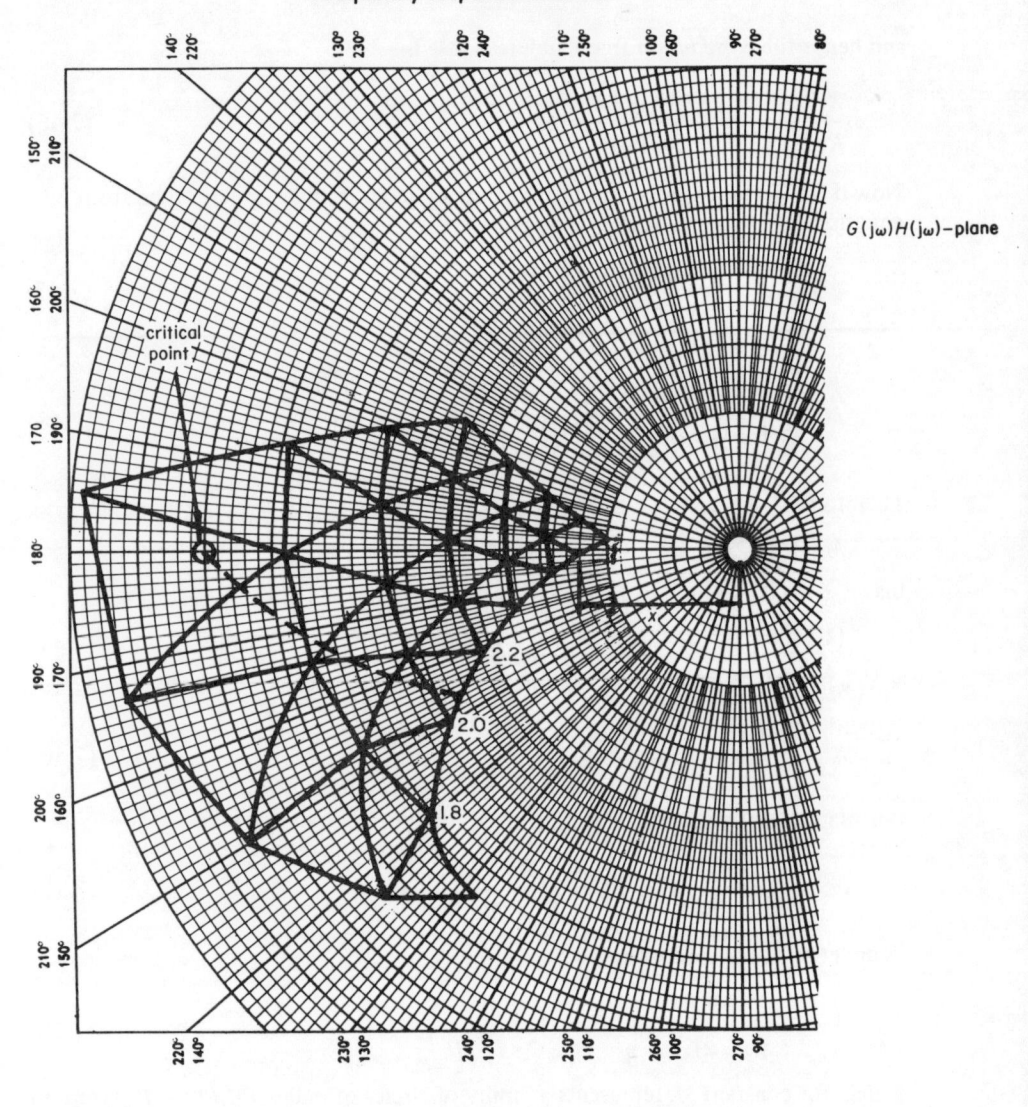

Figure 7.41

M and N Circles

For unity-feedback systems or equivalent unity-feedback systems with forward transference $KG(s)$, the closed-loop transfer function is

$$\frac{X(s)}{R(s)} = \frac{KG(s)}{1 + KG(s)} \tag{7.45}$$

Introduction to Control Theory

and hence the *closed-loop* frequency response is

$$\frac{X(j\omega)}{R(j\omega)} = \frac{KG(j\omega)}{1 + KG(j\omega)} = Me^{j\gamma} \qquad (7.46)$$

Now if $KG(j\omega)$ is expressed in terms of cartesian coordinates $x + jy$, then from equation 7.46

$$Me^{j\gamma} = \frac{x + jy}{1 + x + jy} \qquad (7.47)$$

and

$$M = \sqrt{\left[\frac{x^2 + y^2}{(1 + x)^2 + y^2}\right]}; \; \gamma = \tan^{-1}\frac{y}{x} - \tan^{-1}\frac{y}{1 + x} \qquad (7.48)$$

Therefore

$$M^2(1 + 2x + x^2 + y^2) = x^2 + y^2$$

that is

$$(1 - M^2)x^2 - 2M^2 x + (1 - M^2)y^2 = M^2 \qquad (7.49)$$

Now when $M = 1$

$$x = -\frac{1}{2} \qquad (7.50)$$

and when $M \neq 1$

$$x^2 - \frac{2M^2}{1 - M^2} x + y^2 = \frac{M^2}{1 - M^2} \qquad (7.51)$$

Completing the square gives

$$\left(x - \frac{M^2}{1 - M^2}\right)^2 + y^2 = \left(\frac{M}{1 - M^2}\right)^2 \qquad (7.52)$$

which, for constant M, represents a family of circles of radius $|M/(1 - M^2)|$ centred at $(x, y) = (M^2/(1 - M^2), 0)$. These are the so-called M circles along any of which the closed-loop gain is constant and equal to M; they are shown in figure 7.42. Note that they are symmetrical about the x axis; for $0 < M < 1$ they lie to the right of the line $x = -1/2$ and for $1 < M < \infty$ to the left.

Also from equation 7.48

$$\tan \gamma = \tan\left[\tan^{-1}\frac{y}{x} - \tan^{-1}\frac{y}{1 + x}\right] \qquad (7.53)$$

$$= \frac{y/x - y/(1 + x)}{1 + y^2/x(1 + x)} = N \qquad (7.54)$$

Frequency-response Methods

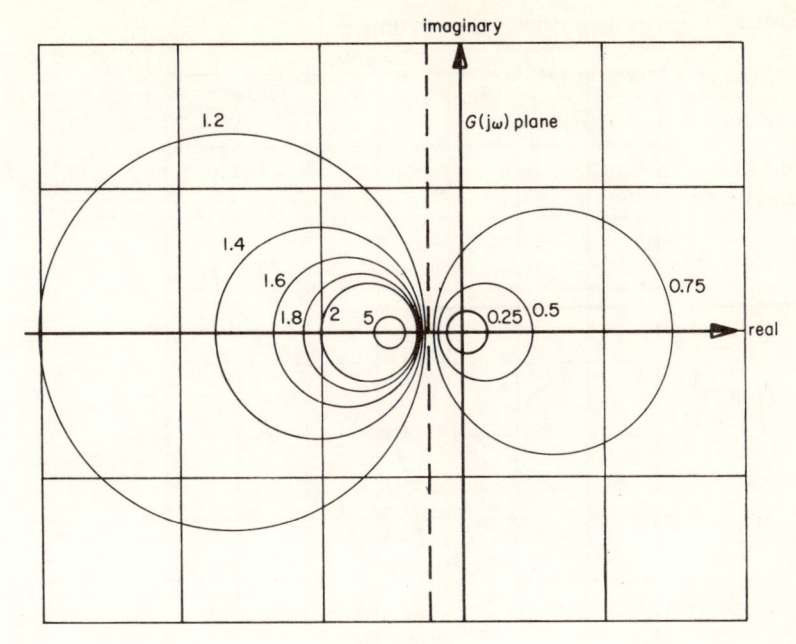

Figure 7.42

Therefore

$$\frac{y}{x} - \frac{y}{1+x} = \left[1 + \frac{y^2}{x(x+1)}\right]N$$

When $N = \tan \gamma = 0$

$$y = 0 \qquad (7.55)$$

and when $N \neq 0$

$$x^2 + x + y^2 - \frac{y}{N} = 0 \qquad (7.56)$$

Completing the square gives

$$\left(x + \frac{1}{2}\right)^2 + \left(y - \frac{1}{2N}\right)^2 = \frac{1}{4}\left(1 + \frac{1}{N^2}\right) \qquad (7.57)$$

which, for constant N, represents a family of circles of radius $(1/2)\sqrt{(1 + 1/N^2)}$ centred at $(x, y) = (-1/2, +1/2N)$. These are the so-called $N (= \tan \gamma)$ circles along any of which the closed-loop phase shift is constant and equal to γ. These are shown in figure 7.43.

It is usual for the M and N circles to be superimposed on either polar or Cartesian coordinates, then known as a *Hall chart* (see figure 7.44), which can be used to convert an open-loop frequency response $KG(j\omega)$ to a closed-loop frequency response

168   Introduction to Control Theory

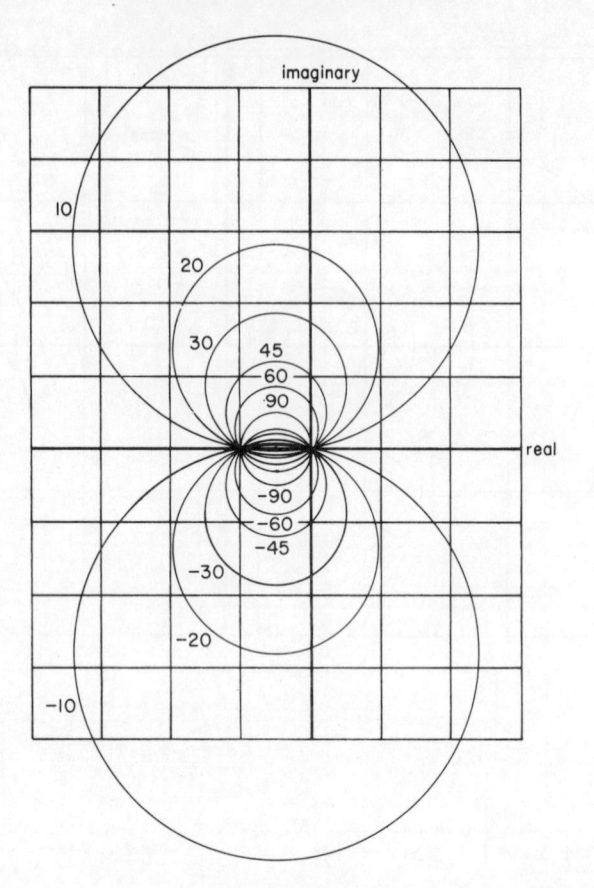

Figure 7.43

by simply reading off the intersections between the $KG(j\omega)$ plot and the superimposed M and N circles. Usually the complete closed-loop frequency response is not required by the designer and it is sufficient to know the maximum closed-loop gain M_p, the resonant frequency ω_0 at which it occurs and also the cut-off frequency ω_c or bandwidth of the system. The latter is defined as the frequency at which the gain drops to -3 db or 0.7. These closed-loop quantities are illustrated in figure 7.45 and the following example will show how they are obtained for a particular system from knowledge of its open-loop frequency response.

Example 7.11
Draw the Nyquist diagram $KG(j\omega)$, for the system of example 7.10 on a Hall chart and obtain the maximum amplitude M_p, the resonant frequency ω_0 and the bandwidth of the closed-loop system.

The Nyquist diagram is shown superimposed on the Hall chart of figure 7.44. To obtain the maximum amplitude M_p establish the M circle that is tangential to $KG(j\omega)$: from the figure this is $M = 1.2$ and therefore $M_p = 1.2$. The resonant frequency $\omega_0 \sim 1.9$ rad/s is the point at which the M_p circle touches.

Frequency-response Methods

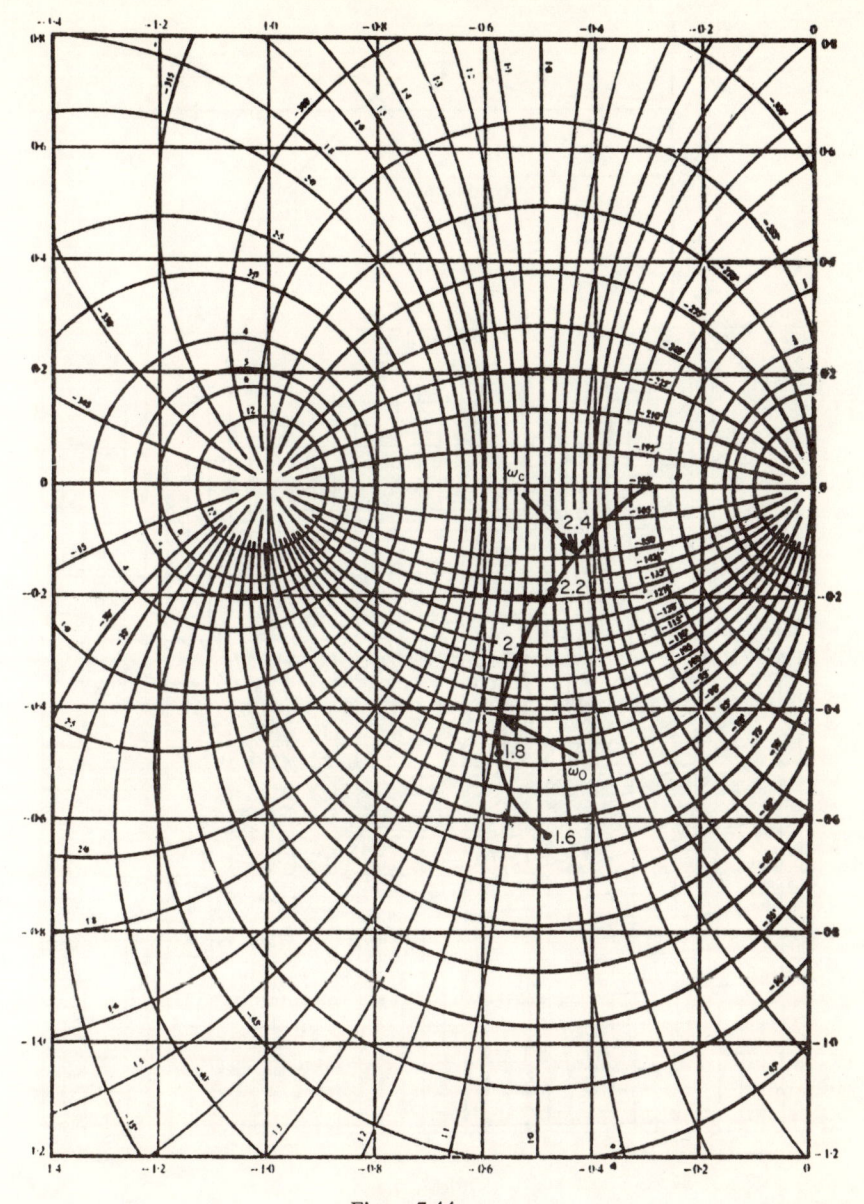

Figure 7.44

To obtain the bandwidth, find the point at which the M circle $M = 0.7$, and the Nyquist diagram intersect, and read off the corresponding frequency, that is, $\omega_c = 2.35$ rad/s: this is the bandwidth.

M and N circles on the Nyquist diagram can be transformed to M and N curves on the Nichols diagram, known as the *Nichols chart* (see figure 7.46), and can be used

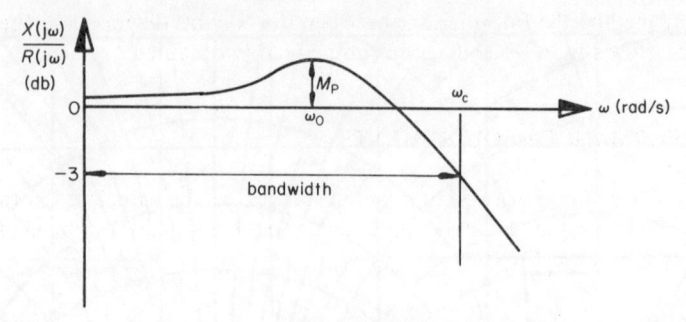

Figure 7.45

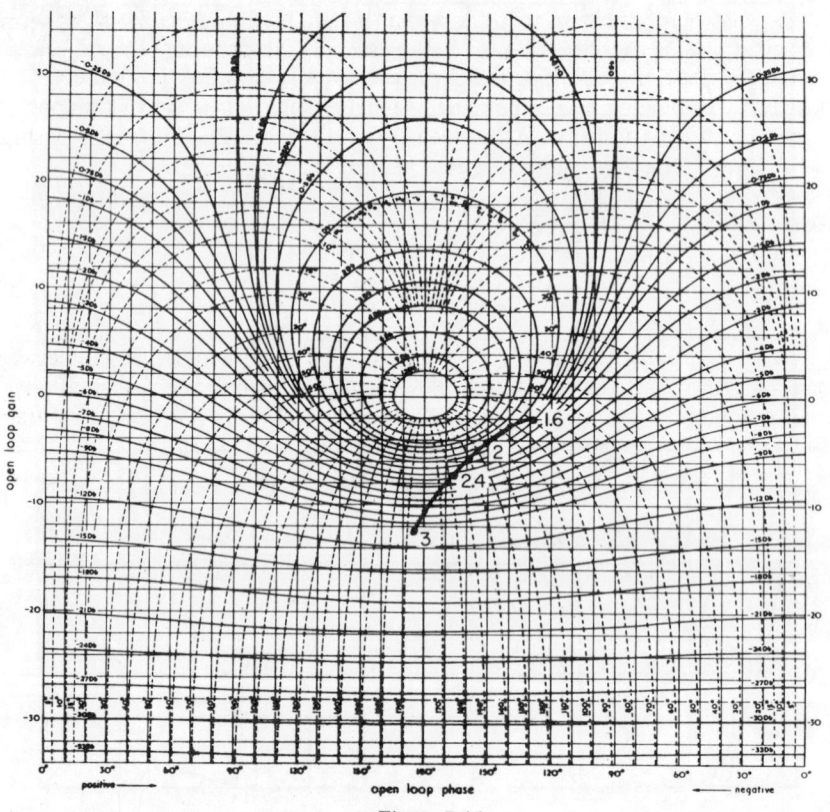

Figure 7.46

in exactly the same way as the Hall chart. In fact, the frequency response of the previous example has been superimposed on figure 7.46 and the reader can verify the values for closed-loop amplitude M_p, resonant frequency and bandwidth. The Nichols chart is often preferred to the Hall chart because asymptotic approximations can be used, gain change is more easily accomplished (simply by moving the 0-db axis) and the scale is more favourable to a wider range of problems.

Finally, by reading the intersections between the Nichols diagram and the M and N curves, the closed-loop Bode diagram can be readily obtained.

7.6 PERFORMANCE REQUIREMENTS

In general a large value of maximum amplitude M_p will be associated in the closed-loop with a small value of the damping factor ξ and it is possible when considering a closed-loop system of the form

$$G'(s) = \frac{\omega_n^2}{s^2 + 2\xi\omega_n s + \omega_n^2} \tag{7.58}$$

that is, a second-order system, to interpret and obtain relationships for M_p in terms of the performance criteria of chapter 5 (the damping factor ξ, the percentage overshoot and the rise time), recalling that a closed-loop system whose response is dominated by a complex pair of poles will generally behave in a similar fashion. Thus, it is possible to design in the frequency domain and, using these relationships, to obtain information relating to transient behaviour.

From equation 7.32 the maximum amplitude exhibited by the second-order system of equation 7.58 is given by

$$M_p = \frac{1}{2\xi\sqrt{1-\xi^2}} \quad (\xi < 0.707) \tag{7.59}$$

and occurs at a resonant frequency $\omega_0 = \omega_n\sqrt{(1-2\xi^2)}$. The relationship between M_p and ξ is shown in figure 7.47a, that between M_p and percentage overshoot in figure 7.47b and that between M_p and $\omega_n t_r$ in figure 7.47c; in this case, percentage overshoot $= 100 \exp[-\xi\pi/\sqrt{(1-\xi^2)}]$ and $t_r = \pi/\omega_n\sqrt{(1-\xi^2)}$.

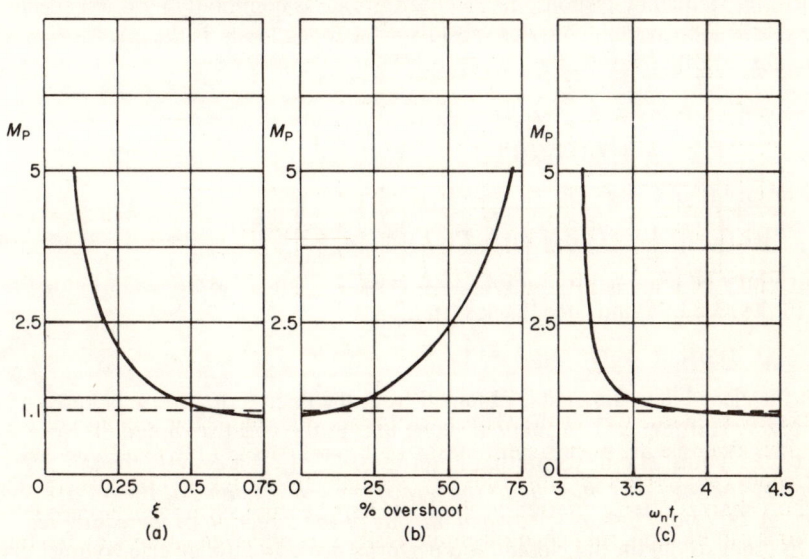

Figure 7.47

Usually a system should have a value of M_p between 1.0 and 1.4, although the actual choice depends on the particular application being considered. For example, if it is a second-order system, it follows from the curves of figure 7.47 that M_p should be chosen not greater than 1.1.

The bandwidth is defined by figure 7.45 and should be chosen so as to be large enough to contain the full frequency content of the input but no larger, in order to minimise any noise problems.

7.7 SYSTEM SENSITIVITY

It was shown in chapter 5 that the sensitivity $S(s)$ of the closed-loop system with unity feedback is defined by

$$S(s) = \frac{\Delta G'(s)/G'(s)}{\Delta G(s)/G(s)} = \frac{1}{1 + KG(s)} \tag{7.60}$$

In the frequency domain for $s = j\omega$

$$|S(j\omega)| = \left|\frac{1}{1 + KG(j\omega)}\right| \tag{7.61}$$

and for small system sensitivity $|1 + KG(j\omega)|$ (that is, the vector $1 + KG(j\omega)$, which is the line drawn from the -1 point to the Nyquist diagram) must be made as large as possible—subject, obviously, to the other performance requirements.

7.8 INVERSE NYQUIST DIAGRAM

A further frequency-response representation that is commonly used, especially when the closed-loop system contains subsidiary feedback loops, is the *inverse Nyquist diagram*—in other words, the polar plot of $G^{-1}(j\omega)$. If

$$G'(j\omega) = \frac{G(j\omega)}{1 + G(j\omega)H(j\omega)} \tag{7.62}$$

the inverse is given by

$$[G'(j\omega)]^{-1} = G^{-1}(j\omega) + H(j\omega) \tag{7.63}$$

which may be obtained by vector addition of G^{-1} and H as shown in figure 7.48.

In the case of a unity-feedback system

$$[G'(j\omega)]^{-1} = G^{-1}(j\omega) + 1 \tag{7.64}$$

and the closed-loop diagram is obtained from the open-loop inverse Nyquist diagram simply by shifting the imaginary axis one unit in the negative direction. Note also that the Nyquist stability criterion in terms of $G^{-1}(j\omega)$ is exactly the same: *for stability the -1 point must lie to the left of the nearest point of the inverse Nyquist diagram $G^{-1}(j\omega)$, $0 \leq \omega < \infty$, when moving in the direction of increasing ω.*

In order to obtain the closed-loop performance using the inverse Nyquist diagram

Frequency-response Methods

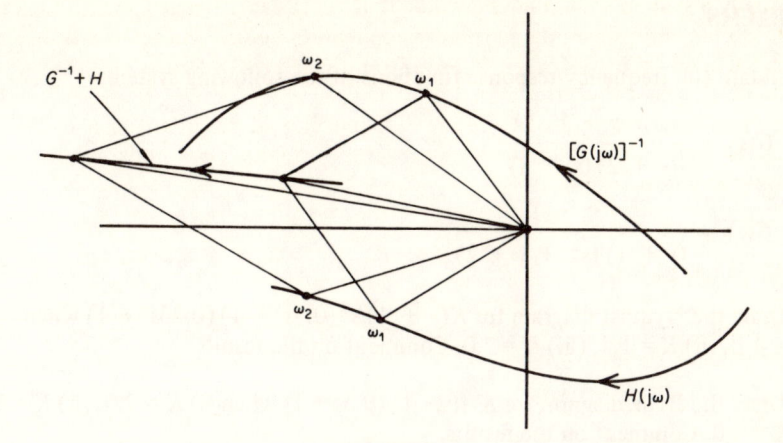

Figure 7.48

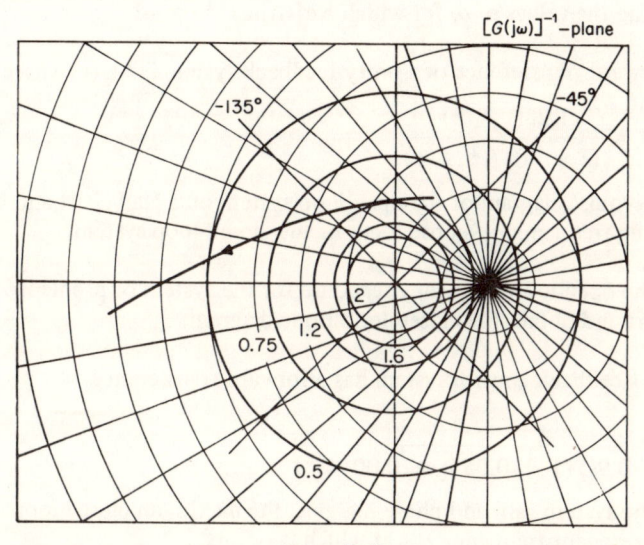

Figure 7.49

it should be recognised that the M circles are simply circles of radius $1/M$ with centre $(-1, 0)$. The inverse Nyquist diagram for example 7.10 is shown in figure 7.49 together with the family of M circles and lines of constant N.

REFERENCES

Murphy, G. J., *Basic Automatic Control Theory* (Van Nostrand, New York, 1966).
Thaler, G. J., and Brown, R. G., *Analysis and Design of Feedback Control Systems* (McGraw-Hill, New York, 1960).

PROBLEMS

7.1 Obtain the frequency-response functions of the following systems

(i) $\quad G(s) = \dfrac{s + 1}{s(s + 2)(s + 3)}$

(ii) $\quad G(s) = \dfrac{3s + 1}{(s + 1)(s^2 + s + 2)}$

7.2 Draw the Nyquist diagram for $K(s + 1)/s^2(0.5s + 1)(0.25s + 1)$ when (i) $K = 1.0$; (ii) $K = 1.5$; (iii) $K = 2.0$. Comment on the results.

7.3 Draw the Bode diagram for $K/s(s + 1)(0.5s + 1)$ when (i) $K = 2.0$; (ii) $K = 3.0$; (iii) $K = 4.0$. Comment on the results.

7.4 Draw the inverse Nyquist diagram for $G(s) = 20\,000/s(s + 20)(s^2 + 10s + 100)$ and determine the values of ω for which $\mathrm{Re}[G(j\omega)^{-1}] = -1$.

7.5 The forward transference of a unity-feedback system is $KG(s)$ where

$$KG(s) = \dfrac{1.25(s + 8)}{(s + 10)(s^2 + 1)}$$

Draw a Nyquist diagram of the open-loop system on a Hall chart and obtain the corresponding frequency-response locus of the closed-loop system.

7.6 Draw the open-loop frequency response for the system of problem 7.5 on a Nichols chart and obtain the closed-loop Bode diagram.

7.7 A unity-feedback control system has a forward transference

$$\dfrac{5.4}{s(1 + 0.9s)(1 + 0.053s + 0.0044s^2)}$$

Obtain the system gain and phase margins, the maximum-closed loop amplitude M_p and the resonant frequency ω_0 at which it occurs.

7.8 A closed-loop system with unity feedback has a forward transference

$$KG(s) = \dfrac{100K}{s(100 + 10s + s^2)(s + 1)}$$

Use Routh to establish the value of K that makes the closed-loop system oscillate continuously with constant amplitude. For a K of half this value find the gain and phase margins of the system and the maximum closed-loop amplitude M_p.

8 Control-system Synthesis

A control system is designed to meet certain performance requirements, which usually include system stability, steady-state accuracy, dynamic accuracy, and insensitivity to parameter changes. These performance requirements can be expressed either in terms of the system transient response (quantities such as the maximum overshoot, rise time, settling time, damping factor and undamped natural frequency of the desired dominant closed-loop poles) or in terms of the system frequency response (quantities such as the phase margin, gain margin, maximum magnification and error coefficients) or both.

Usually, if the performance requirements are expressed in terms of the system transient response, the root-locus method is chosen to design the system and, if they are expressed in terms of the frequency response, either the Bode diagram or the Nyquist diagram (Hall chart or Nichols chart) is used, depending on whether the phase or gain margins are specified or the maximum magnification is specified.

In chapters 6 and 7 it was shown how a feedback control system with a variable gain controller could be analysed (that is, it was shown how the variation in gain affects the behaviour of the system) and also designed (that is, how the gain should be chosen to satisfy some performance requirement). For some control systems, variation in gain, or gain compensation, will not satisfy or meet the stated performance; for example, the gain required to meet the specified error coefficients may be too high, leading to a system with a highly oscillatory response. In other cases, where the system is open-loop unstable, it is impossible to stabilise the closed-loop system using gain compensation only. Thus other forms of compensation have to be introduced into the closed-loop system.

In chapter 5, the addition of various compensators was considered and their effect upon the steady-state and transient response of closed-loop control systems was analysed. These compensators included derivative action, integral action, and velocity feedback (see figure 8.1); all consist of variable frequency-dependent terms in addition to the variable gain terms.

Thus compensators may be added in both the forward path and feedback path of a closed-loop system; they are known, respectively, as series compensators and parallel or feedback compensators (see figure 8.2).

The types of compensator are many and varied and may be classified theoretically as gain, integral, differential, lead and lag. In practice, however, pure different-

176 Introduction to Control Theory

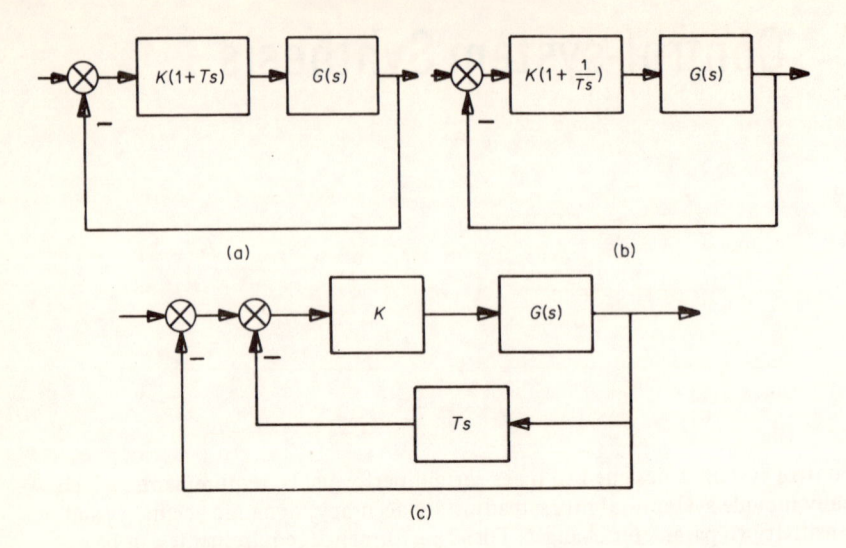

Figure 8.1 (a) Derivative action; (b) integral action; (c) velocity feedback

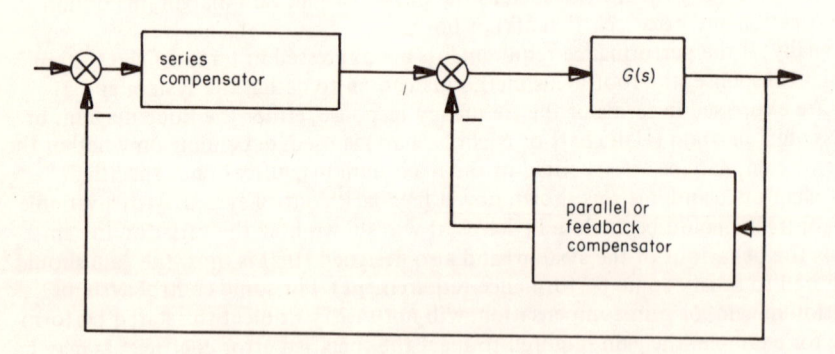

Figure 8.2

iation cannot be achieved (because of problems of realisation and noise) and, as was shown in chapter 7, lead compensation is used as a practical alternative, since it is an acceptable approximation to differentiation over a particular bandwidth of frequencies. Also, lag compensation is used as an approximation to integral action.

The aims of this chapter are three: to discuss the various forms of compensation; to analyse their effect upon the behaviour of the closed-loop system; to show how to choose and design a compensator so as to achieve a desired system performance. This is known as *control-system synthesis*. It is considered in the *s*-plane using the root-locus method and subsequently in the $G(j\omega)H(j\omega)$ plane using the various frequency-response methods. Following each design, the transient response of the compensated closed-loop control system is obtained to confirm that design.

Many of the results were obtained using the graphics terminal.

8.1 SERIES COMPENSATION

There are four types of series compensator to be considered here—variable-gain, lead-network, lag-network and lead–lag-network compensators; they are shown in figure 8.3.

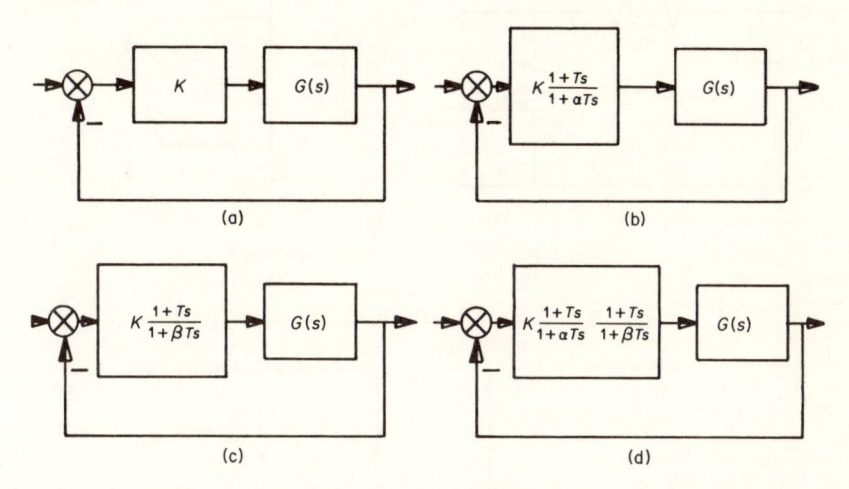

Figure 8.3 (a) Variable-gain compensator; (b) lead network ($\alpha < 1$); (c) lag network ($\beta > 1$); (d) lead–lag network

Variable-gain Compensation

The effect of changing the gain in the forward path of a closed-loop control system has been discussed in chapters 5, 6 and 7 and will not be discussed further here, although comparisions will be made to indicate improvement in performance achieved by the use of the other compensators relative to gain compensation only. It is stressed that, before considering the use of other compensators, an attempt should be made to obtain the required system performance by means of gain compensation only, mainly on the grounds of simplicity and economy.

Lead-network Compensation

Lead-network compensation is characterised by the transfer function

$$G_c(s) = \frac{\alpha(Ts + 1)}{\alpha Ts + 1} = \frac{s + 1/T}{s + 1/\alpha T} \quad (\alpha < 1) \tag{8.1}$$

which can be realised by the RC network in figure 8.4a or the mechanical system in figure 8.4b. The pole–zero plot and the Nyquist diagram of the lead network are given in figures 8.5a and b, respectively. Figure 8.5a shows that the zero is nearer the origin than the pole and figure 8.5b indicates a Nyquist diagram consisting of a semi-circle with its base on the real positive axis, positioned in the first quadrant. It follows that the phase associated with the lead network is leading for all frequencies.

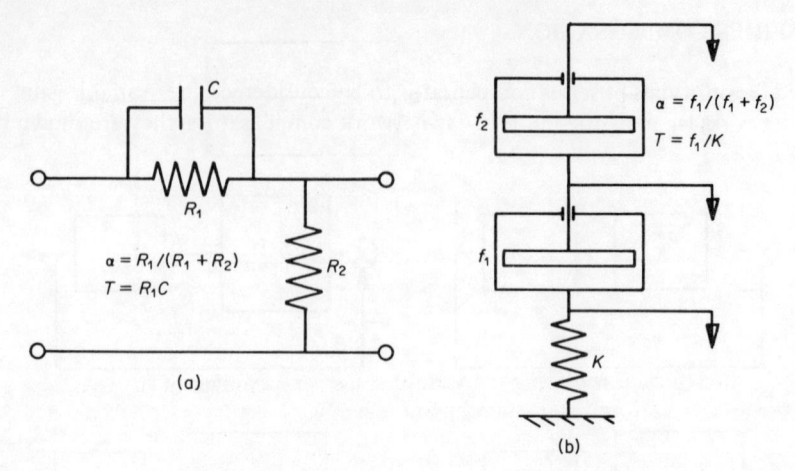

Figure 8.4 (a) RC network; (b) mechanical system

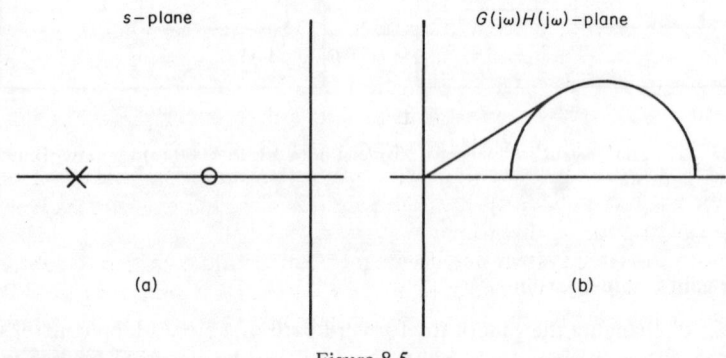

Figure 8.5

From example 7.3 the maximum phase-angle lead is given by

$$\phi_{max} = \sin^{-1} \frac{1 - \alpha}{1 + \alpha} \tag{8.2}$$

(see table 8.1) and occurs at a frequency $\omega = 1/T\sqrt{\alpha}$ with magnitude $1/\sqrt{\alpha}$.

Table 8.1

α	ϕ_{max}
0.1	55°
0.2	42°
0.3	33°
0.4	25°
0.5	19°

From considerations of practical realisation, α is never chosen to be smaller than 0.1 and so the maximum phase-angle lead possible with this type of network is 55°.

Control-system Synthesis

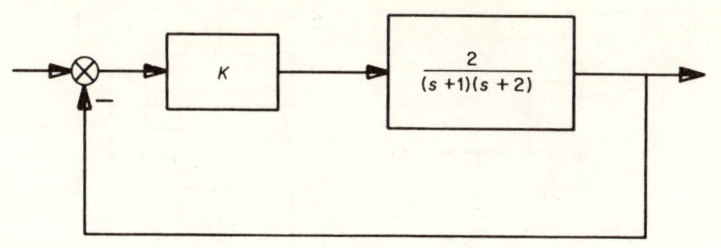

Figure 8.6

Example 8.1

Investigate the effect of replacing the variable-gain compensator of figure 8.6 by a lead compensator with transfer function

$$G_c(s) = K_1 \frac{s + 3}{s + 4.2}$$

The root-locus diagram is given in figure 8.7a, where it is seen that K is chosen so that the dominant poles subtend an angle of $60°$ with the negative real axis; the resulting transient response is given in figure 8.7b.

Thus with $K = 3.5$ the system exhibits a suitable transient response but also a fairly large steady-state error. With the variable-gain compensator the steady-state error can be reduced by increasing the gain but the transient response will deteriorate, since the angle subtended by the dominant poles will increase and lead to a reduction in damping factor, and so on.

For the compensated system, the dominant poles are again chosen to subtend an angle of $60°$ with the negative real axis, giving $K_1 = 7.0$. It is seen in figure 8.7b that the compensated system has an improved transient response and also a reduced steady-state error.

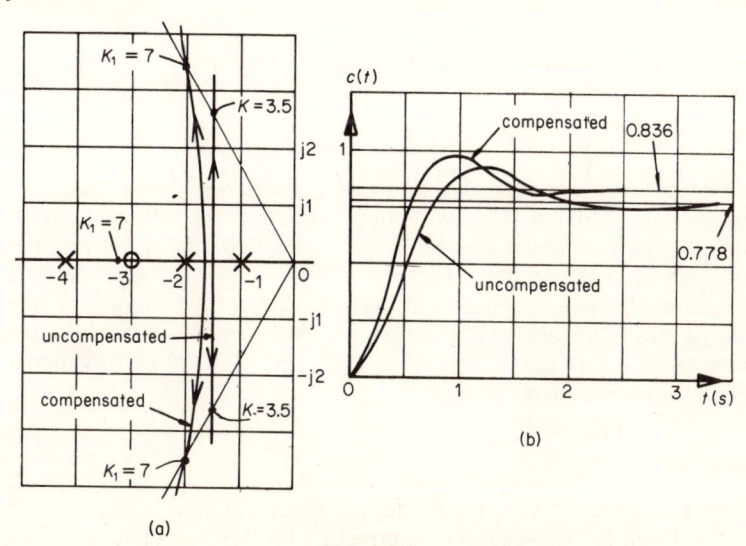

Figure 8.7

180   Introduction to Control Theory

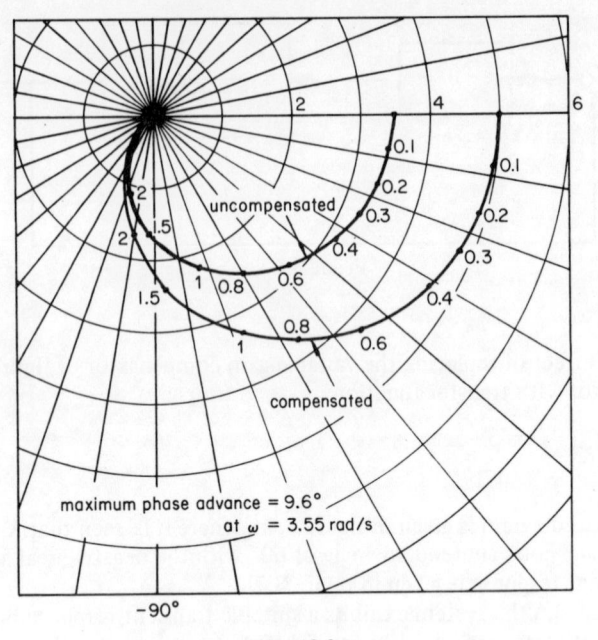

Figure 8.8

Figure 8.8 shows the Nyquist diagrams for the two systems. Comparison of the Nyquist diagrams shows clearly that the compensator is leading and that it is more effective at high frequencies. This confirms that the lead network allows the system gain to be increased without adversely affecting stability and transient response (high frequency) resulting in a reduction in steady-state error (low frequency).

Lag-network Compensation

Lag-network compensation is characterised by the transfer function

$$G_c(s) = \frac{Ts + 1}{\beta Ts + 1} = \frac{1}{\beta}\left(\frac{s + 1/T}{s + 1/\beta T}\right) \quad (\beta > 1) \tag{8.3}$$

which can be realised by the *RC* network in figure 8.9a or the mechanical system in figure 8.9b. The pole-zero plot and the Nyquist diagram of the lag network are given in figures 8.10a and b, respectively. Figure 8.10a shows the pole nearer to the origin than the zero and figure 8.10b indicates a Nyquist diagram consisting of a semi-circle with its base on the real axis, positioned in the fourth quadrant. It follows that the phase associated with the lag network is lagging for all frequencies. From example 7.3 the maximum phase-angle lag is

$$\phi_{max} = \sin^{-1}\frac{1 - \beta}{1 + \beta} \tag{8.4}$$

and occurs at a frequency $\omega = 1/T\sqrt{\beta}$ with magnitude $1/\sqrt{\beta}$.

Control-system Synthesis

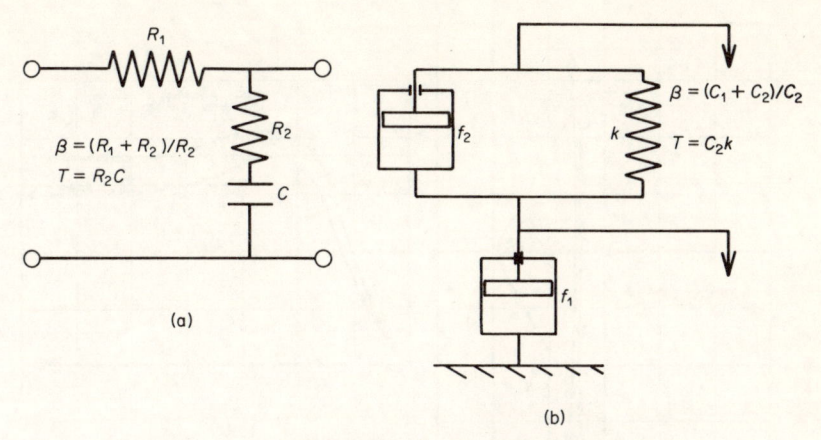

Figure 8.9 (a) *RC* network; (b) mechanical system

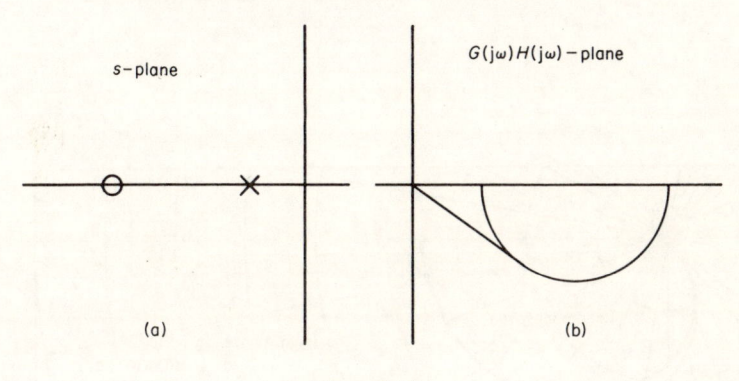

Figure 8.10

Example 8.2
Investigate the effect of replacing the variable-gain compensator of figure 8.6 by the lag compensator with transfer function

$$G_c(s) = K_1 \frac{s + 0.34}{s + 0.1}$$

The root-locus and transient response with gain $K = 3.5$ of figure 8.6 are shown in figures 8.7a and b, respectively and are repeated for the sake of comparison in figures 8.11a and b.

For the compensated system, the dominant poles are again chosen to subtend an angle of 60° with the negative real axis, giving $K_1 = 3.175$. It is seen in figure 8.11b that the compensated system has a slower transient response but a much-reduced steady-state error.

Introduction to Control Theory

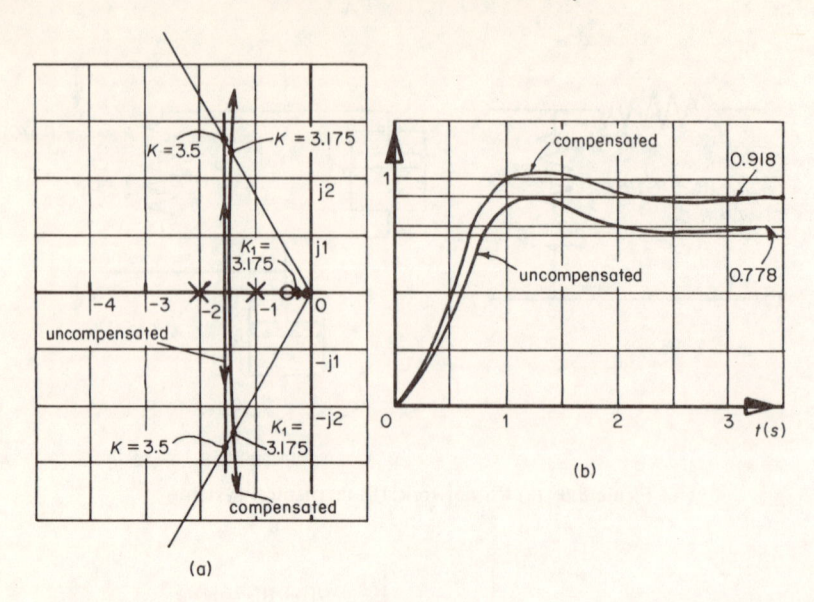

Figure 8.11

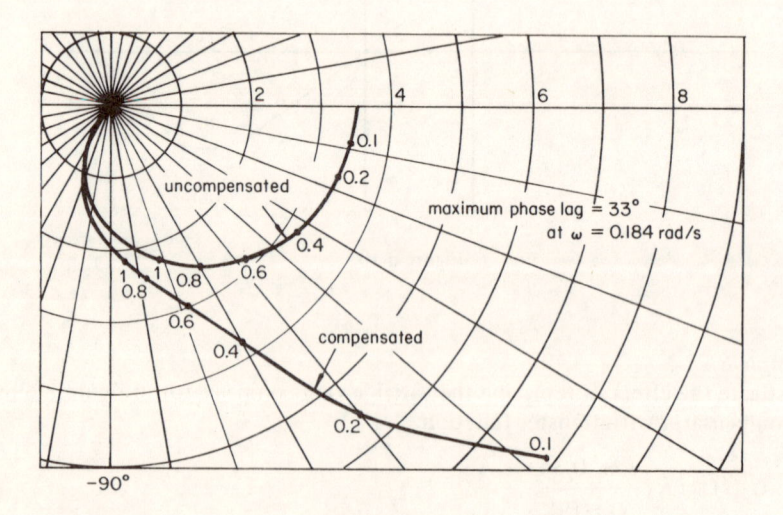

Figure 8.12

Figure 8.12 shows the Nyquist diagram for the two systems. Comparison clearly indicates that the compensator is lagging and that it is more effective at low frequencies. Thus, provided that the pole-zero locations are correctly chosen, the lag network will reduce the steady-state error whilst not seriously affecting the transient response.

Control-system Synthesis   183

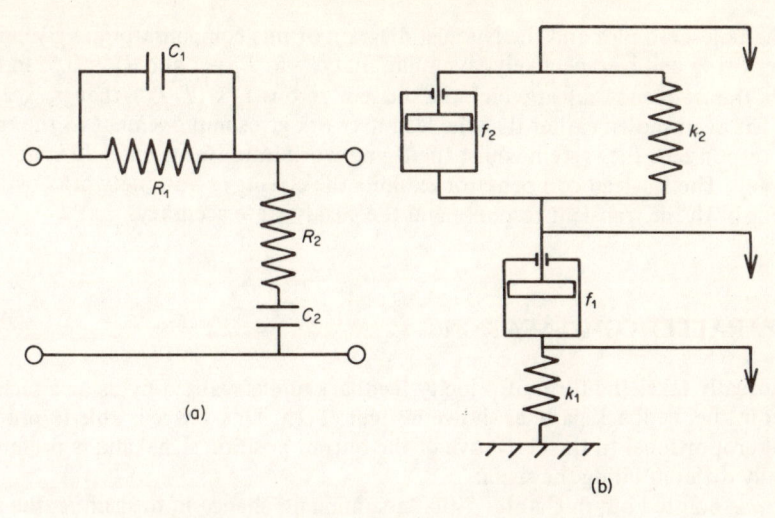

Figure 8.13 (a) RC network; (b) mechanical action

Lag-Lead-network Compensation

Lag-lead compensation may be achieved by a series combination of a lag network and a lead network or alternatively, by a single network as in figure 8.13a or by the mechanical system of figure 8.13b. The appropriate form of the transfer function is

$$G_c(s) = \left(\frac{1 + T_1 s}{1 + \alpha T_1 s}\right)\left[\frac{1 + T_2 s}{1 + (T_2/\alpha)s}\right] \qquad (\alpha > 1) \tag{8.5}$$

in which the first term produces the lag effect and the second term the lead effect.

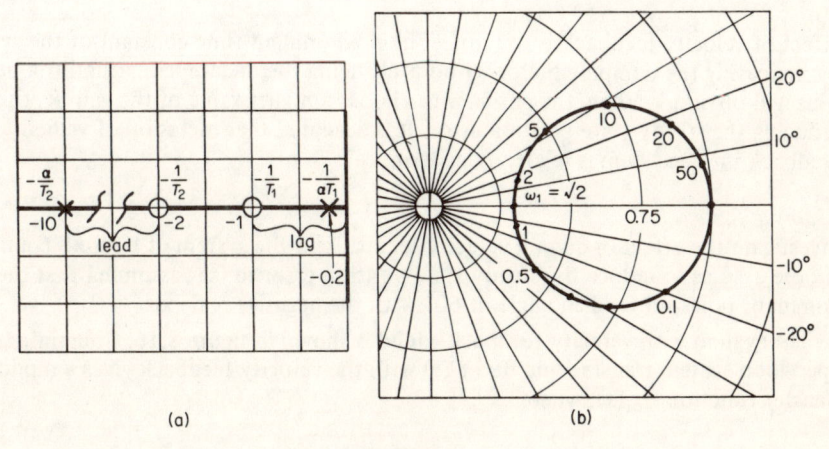

Figure 8.14

The pole–zero plot and the Nyquist diagram of this compensator are given in figures 8.14a and b respectively, assuming that $\alpha = 5$, $T_1 = 1$ and $T_2 = 0.5$. In figure 8.14b, the frequency ω_1 at which the phase is zero is $1/\sqrt{(T_1 T_2)}$, that is, $\sqrt{2}$ rad/s.

It has been shown earlier that the lead network gives improvement to the transient response of the system whilst the lag network improves the steady-state accuracy. The lag–lead compensator exhibits the effects of both networks by improving both the transient response and the steady-state accuracy.

8.2 PARALLEL COMPENSATION

This usually takes the form of velocity feedback and is realised by using a tachometer in the feedback path, as shown in figure 8.15. This device is able to produce a signal proportional to the derivative of the output position signal and is preferred to actually differentiating the signal.

It was pointed out in chapter 5 that assuming no change in the gain K, the addition of velocity feedback has a stabilising effect upon a closed-loop control system; the

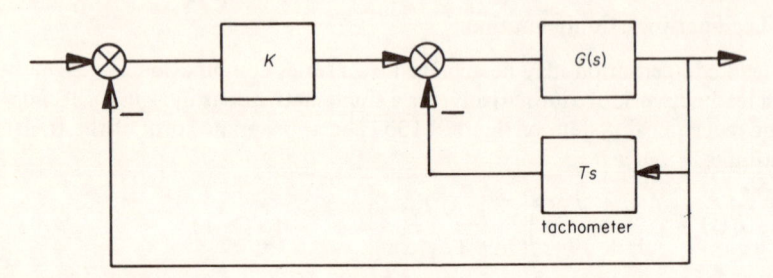

Figure 8.15

effect of velocity feedback is to reduce the predominant time constant of the system by increasing the damping factor without changing the undamped natural frequency. This will obviously allow the designer to choose a higher value of the gain K, thus reducing the steady-state position error. It is accepted that inclusion of velocity feedback into a system is relatively economical.

Example 8.3
Investigate the effect of adding velocity feedback to the system of figure 8.6 and choose T so as to reduce the steady-state position error to 0.1, assuming that the dominant poles subtend an angle of $60°$ with the negative real axis.

The system with velocity feedback added is shown in figure 8.16. Combining the open-loop-system transfer function $G(s)$ with the velocity feedback gives a modified transfer function $G_m(s)$, where

$$G_m(s) = \frac{2}{s^2 + (3 + 2T)s + 2}$$

Control-system Synthesis

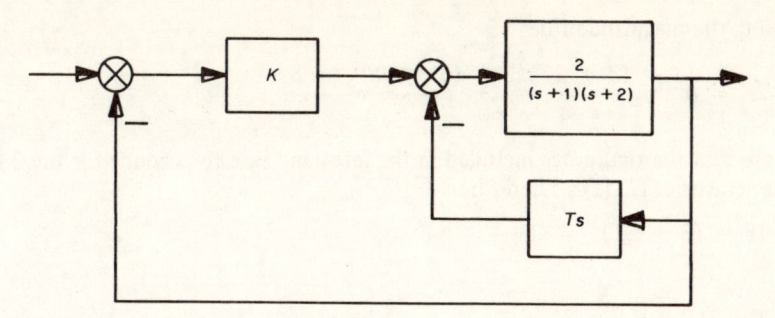

Figure 8.16

and from chapter 5, the steady-state position error following a unity reference signal is given by

$$\epsilon_{ss} = \lim_{s \to 0} \frac{1}{1 + KG_m(s)} = \frac{1}{1 + K}$$

and, since $\epsilon_{ss} = 0.1$ (given), $K = 9$.

The problem now is to choose a suitable value of T such that the dominant poles of the closed-loop system subtend an angle of $60°$ with the negative real axis.

Figure 8.17 shows the positions of the open-loop poles of the modified transfer function in terms of T. It is now a fairly straightforward calculation using the magnitude rule to obtain a relationship between the chosen value of K and the unknown T.

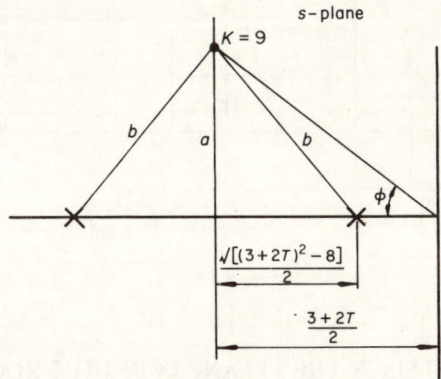

Figure 8.17

From figure 8.17

$$a = \left(\frac{3 + 2T}{2}\right)\tan\phi = \left(\frac{3 + 2T}{2}\right)\sqrt{3}$$

Therefore

$$b = \sqrt{\left[\frac{(3 + 2T)^2 3}{4} + \frac{(3 + 2T)^2 - 8}{4}\right]}$$

and using the magnitude rule

$$2K = b^2 = \frac{(3 + 2T)^2 3}{4} + \frac{(3 + 2T)^2 - 8}{4}$$

(Note that the figure 2 is included in the left-hand side to account for the 2 in the numerator of $G_m(s)$.) Therefore

$$18 = (3 + 2T)^2 - 2$$

and

$$T = 0.736 \text{ s}$$

Using the given value of gain and the calculated value of T the root-locus diagram and the transient response may both be drawn. These are shown in figures 8.18a and b, respectively.

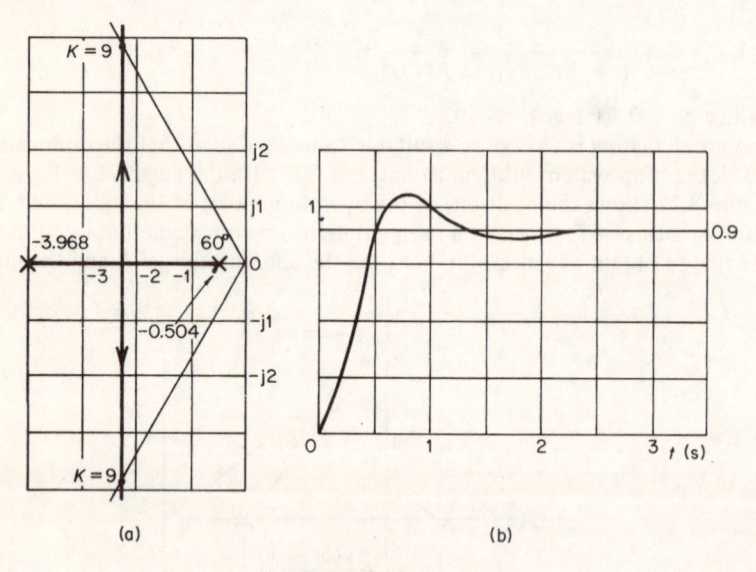

Figure 8.18

8.3 SYSTEM SYNTHESIS IN THE s-PLANE USING THE ROOT-LOCUS METHOD

System synthesis in terms of the root-locus method is essentially based on the use of the s-plane for the evaluation of various compensators. Thus, given the fixed elements of the closed-loop control system, that is, $G(s)$, the design problem is to introduce compensatory open-loop zeros and poles such that the root locus passes through certain points in the s-plane and thereby specific closed-loop-system poles are achieved. These points should be chosen to meet such specifications as the steady-state requirements, bandwidth, rise time, percentage overshoot, frequency of oscillation and settling time. Note that the specifications are all related to the system transient response.

Control-system Synthesis

It is essential to understand the mechanism by which the addition of compensatory poles and zeros to the fixed poles and zeros changes the root-locus shapes. Such an understanding can be obtained by the examination of basic types of root locus, starting with a second-order system and building up to a fourth-order system. The arguments advanced can be extended to higher-order systems by zoning the s-plane into smaller regions and sketching the locus according to the one corresponding to the basic type of locus.

Second-order Systems

Figure 8.19 shows the effect on the root-locus diagram of adding one compensating zero and pole to a second-order system with both poles on the negative real axis. Figures 8.19a, b and c correspond to the addition of a lead network in the cases when the zero is between the fixed poles, immediately to the left of the fixed poles

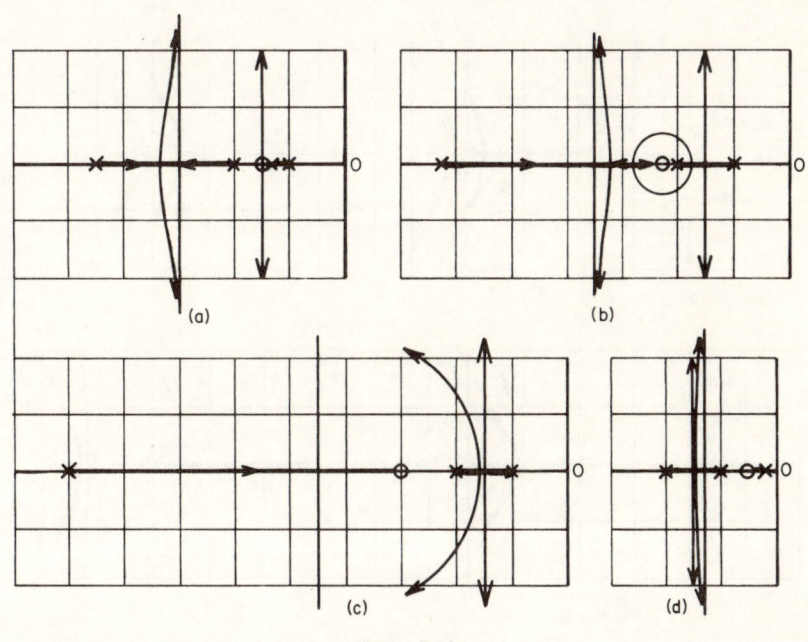

Figure 8.19

and to the left of the fixed poles, respectively. Figure 8.19d corresponds to the addition of a lag network positioned near the origin and to the right of the fixed poles and figure 8.20 shows the effect of adding a lead network to an unstable second-order system, indicating clearly that by choosing a gain greater than K_1 the system can be stabilised.

Third-order Systems

Figure 8.21 indicates the result of adding a lead network to a third-order system with one pole at the origin.

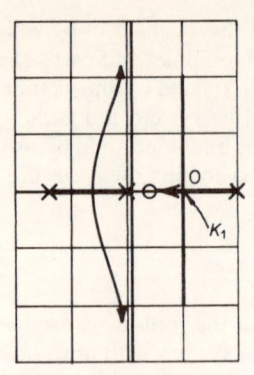

Figure 8.20

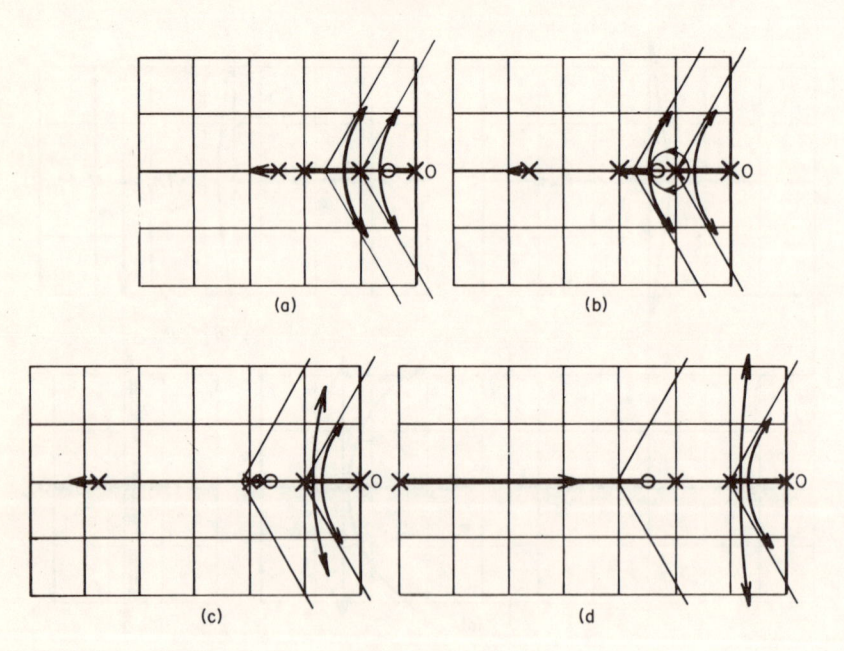

Figure 8.21

Fourth-order systems

Figure 8.22 shows the effect of adding a lead network to a fourth-order system.

In the case of third and fourth-order systems with all poles real, negative and non-zero, the effect of adding a properly designed lag network (see figure 8.19d) is to move the dominant arms of the root-locus plot slightly to the right. As an exercise, the reader should sketch the effect of adding compensatory poles and zeros to third- and fourth-order systems with fixed complex poles.

Control-system Synthesis 189

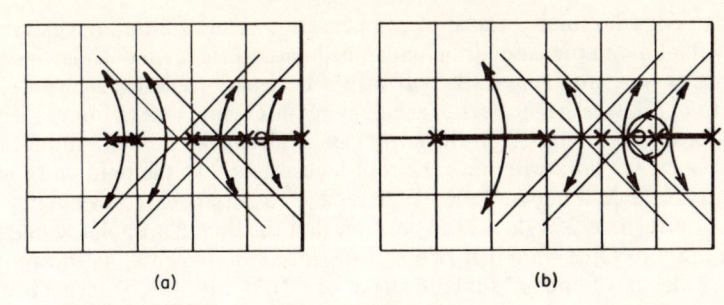

Figure 8.22

System synthesis in the s-plane is to be illustrated first by an example that requires the design of both lead and lag compensation in order to achieve the performance requirements, and secondly (ensuing from the first), a generalised design procedure will be outlined.

Example 8.4
It is required to design a compensator for the system of figure 8.23 such that the dominant poles of the closed-loop system subtend an angle of 60° with the negative real axis, the undamped natural frequency is 5 rad/s and the position error constant is not less than 19.

Figure 8.23

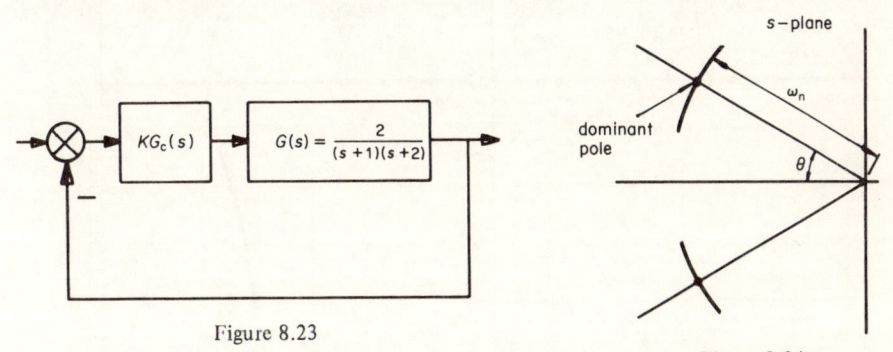

Figure 8.24

Initially it is necessary to interpret the meaning of the performance requirements in the s-plane. The dominant poles must subtend an angle of 60° with the negative real axis and the undamped natural frequency ω_n is required to be 5 rad/s: these requirements are indicated in figure 8.24. From chapter 5, the position error constant K_p is defined as

$$K_p = \lim_{s \to 0} [KG_c(s)G(s)]$$

and is required to be not less than 19.

From example 8.1 and figure 8.7a it is evident that gain compensation only, that is, $G_c(s) = 1$, will not satisfy the first two performance requirements and so it is necessary to consider further forms of compensation. The addition of a suitable lead network will allow an increase in the undamped natural frequency while the damping factor remains constant. Thus initially it will be necessary to design such a network and then to check whether the static accuracy requirement is met; if not, a lag network will have to be designed. In the latter case, it should be borne in mind that addition of a lag network will move the root-locus slightly to the right and some amendment of the lead network may be necessary to correct this movement.

Point A on figure 8.25a shows the position that the dominant poles would have to take in order to satisfy the first two performance requirements. As shown in figure 8.25a, the poles at -1 and -2 subtend angles of $-109°$ and $-96.6°$ respectively, and so, if the root-locus is to pass through point A, the lead network must contribute an angle α, where $\alpha = -180 + 109 + 96.6 = 25.6°$.

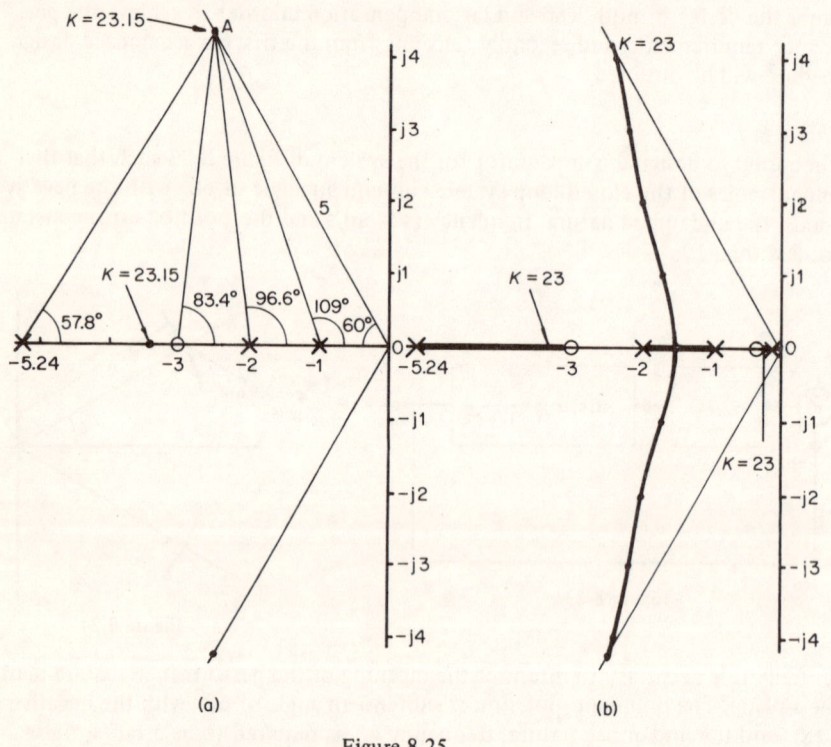

Figure 8.25

Now, by referring to figure 8.19c, choose the zero of the lead network to be positioned at, say, -3, subtending an angle of $83.4°$; then the added pole will have to subtend an angle of $25.6 - 83.4 = -57.8°$ and so will have to be positioned at -5.24.

With this design there will be two branches of the root-locus on the real axis: one branch between the poles at -1 and -2, which will break away from the real axis to form the dominant branches of the locus; and one branch between the pole at -5.24 and the zero at -3. Thus, one of the poles of the closed-loop system will

Control-system Synthesis

always be less than -3 whatever the value of gain and will not severely effect the transient response of the closed-loop system: this is the reason for choosing the zero well to the left of the open-loop poles.

Since the locus will pass through point A, the dominant closed-loop poles will be $-2.5 \pm j4.35$ and the third pole will be $-8.24 + 2 \times 2.5 = -3.24$, since the sum of the roots will be independent of gain. The gain at this point can be determined from the magnitude rule

$$K = \frac{(5.24 - 3.24)(3.24 - 2)(3.24 - 1)}{3.24 - 3} = 23.15$$

It is now required to check the static accuracy requirement; in fact

$$K_p = \lim_{s \to 0} \left\{ 23.15 \frac{s + 3}{s + 5.24} \frac{1}{(s + 1)(s + 2)} \right\} = 6.6$$

which is far below that required. Hence a lag network must be inserted into the system and this is chosen in the following way.

First choose the pole of the network near the origin, say, at -0.1; then the position of the zero is chosen to satisfy the static accuracy requirement

$$\text{Required } K_p = 19 = 6.6 \lim_{s \to 0} \left\{ \frac{s + \alpha}{s + 0.1} \right\}$$

Therefore

$$\alpha = \frac{19 \times 0.1}{6.6} = 0.288 \approx 0.3$$

The angle subtended at point A by the pole at -0.1 is $-118.9°$ and that by the zero at -0.3 is $116.8°$; the difference is $-2.1°$, which will not seriously affect the earlier design.

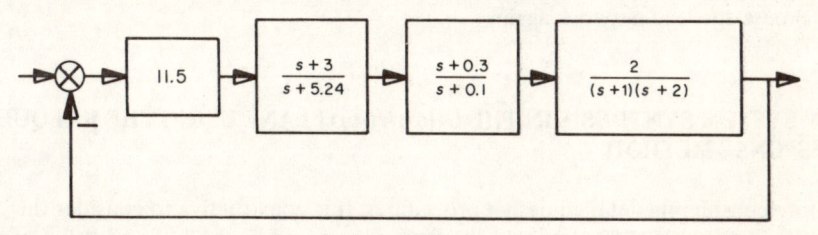

Figure 8.26

The resulting root-locus diagram is given in figure 8.25b, where it is seen that the dominant poles subtend the required $60°$ and the undamped natural frequency is 4.9 rad/s. The block diagram of the closed-loop control system is shown in figure 8.26 and the resulting transient response in figure 8.27. This is compared with the open-loop response.

From an understanding of the solution to the above problem it is possible to outline a generalised procedure for system synthesis in the s-plane using the root-locus method.

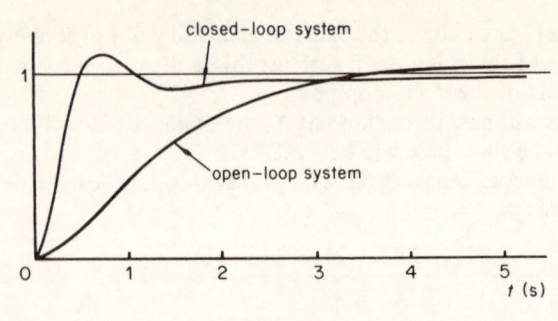

Figure 8.27

(1) Interpret the performance requirements in terms of the required position of the closed-loop dominant poles in the s-plane.
(2) Design a lead network that will cause the root-locus to pass through the required dominant poles; position the zero first and then the pole, and bear in mind the limitations of practical realisation. Calculate the resulting value of the gain K.
(3) Check to see if this design satisfies the required steady-state performance, that is, K_p, K_v or K_a.
(4) If not, design a suitable lag network, first positioning the pole closer than the zero to the origin to satisfy a particular K_p, K_v or K_a.
(5) Re-draw the root locus and check that the introduction of the lag network does not seriously affect the shape of the locus. Calculate the resulting gain and the position of the closed-loop poles.
(6) Calculate the resulting transient response and check that all performance requirements are met.
(7) Some trial-and-error design may be required; that is, step (4) may seriously affect the shape of the locus and hence the transient response; and it may be necessary either to choose the lag network again or to go back to step (2) and choose the lead network again.

8.4 SYSTEM SYNTHESIS IN THE $G(j\omega)H(j\omega)$ PLANE USING THE FREQUENCY-RESPONSE METHOD

Before considering detailed design procedures, it is constructive to consider the results of example 8.4 in terms of the Bode diagram of figure 8.28. Addition of the lead network allows the gain of the system to be increased while allowing an increase in the phase margin; thus the stability of the system and the transient response are improved and the steady-state error is decreased, although the bandwidth of the system (that is, the frequency at which the gain drops to -3 db or 0.7) has been increased. Addition of the lag network is seen not to affect the frequency response in the vicinity of the gain cross-over frequency (that is, the frequency at which the gain is 0 db or 1.0) and thus there is no effect on the transient response; however, the frequency response is affected at frequencies below ~ 1 rad/s since the increase in the system gain in this vicinity results in a reduced steady-state error.

Control-system Synthesis

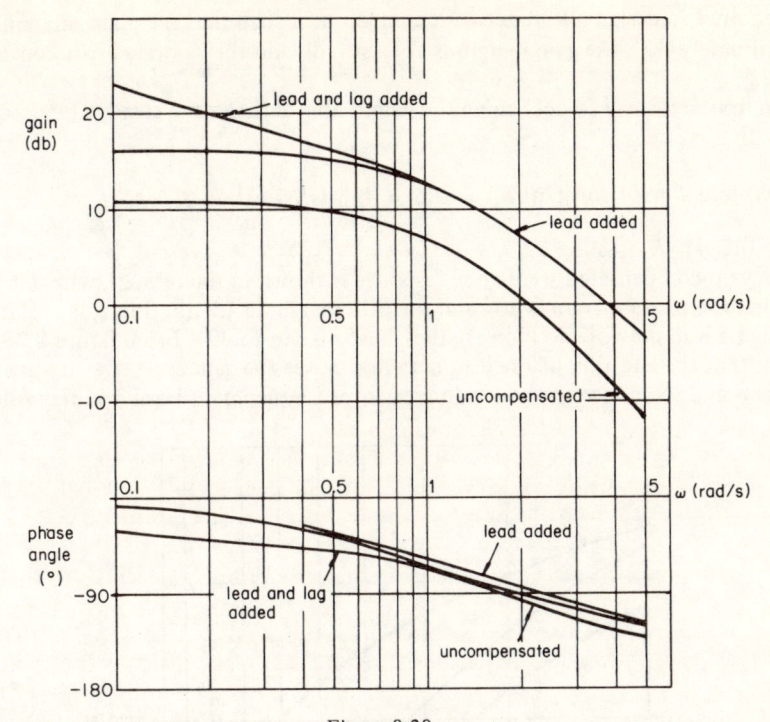

Figure 8.28

The example amply illustrates three particular design requirements.

(1) The slope of the amplitude-log ω curve in the vicinity of the gain cross-over frequency should be approximately −20 db/decade in order to ensure an adequate phase margin, and hence a stable system. This statement can be verified by consideration of Bode's first theorem.
(2) The lead network should be most effective in the vicinity of the required gain cross-over frequency.
(3) The lag network should be most effective at frequencies a decade less than the required gain cross-over frequency.

System synthesis in the $G(j\omega)H(j\omega)$ plane is best illustrated by considering a specific example from which generalised design procedures will be evident. It should be borne in mind that, once a suitable compensator has been designed in the $G(j\omega)H(j\omega)$ plane, the resulting transient response of the closed-loop system should be computed to verify the system behaviour. Consider again the example already considered in the previous two chapters.

Example 8.5
A unity-feedback system has a forward transference

$$KG(s) = \frac{K(s + 1)}{s(0.5s + 1)(0.33s + 1)(0.25s + 1)}$$

Introduction to Control Theory

It is required to design a lead-network compensator such that the phase margin is approximately 40°, the gain margin is at least 6 db and the velocity error constant K_v is 3.5.

It is first necessary to determine the gain K that satisfies the steady-state requirements, that is

$$\text{Velocity error constant } K_v = 3.5 = \lim_{s \to 0} [sKG(s)] = K$$

and so the gain K must be 3.5.

The gain-compensated frequency response is shown in figure 8.29, where it is seen that the phase margin is 20° and the gain margin is 4.7 db; thus it is required to design a lead network to increase the phase margin to 40°. From figure 8.28 it is evident that the addition of the lead network moves the gain cross-over frequency to the right, and so choose the new gain cross-over frequency to be at 5 rad/s initially.

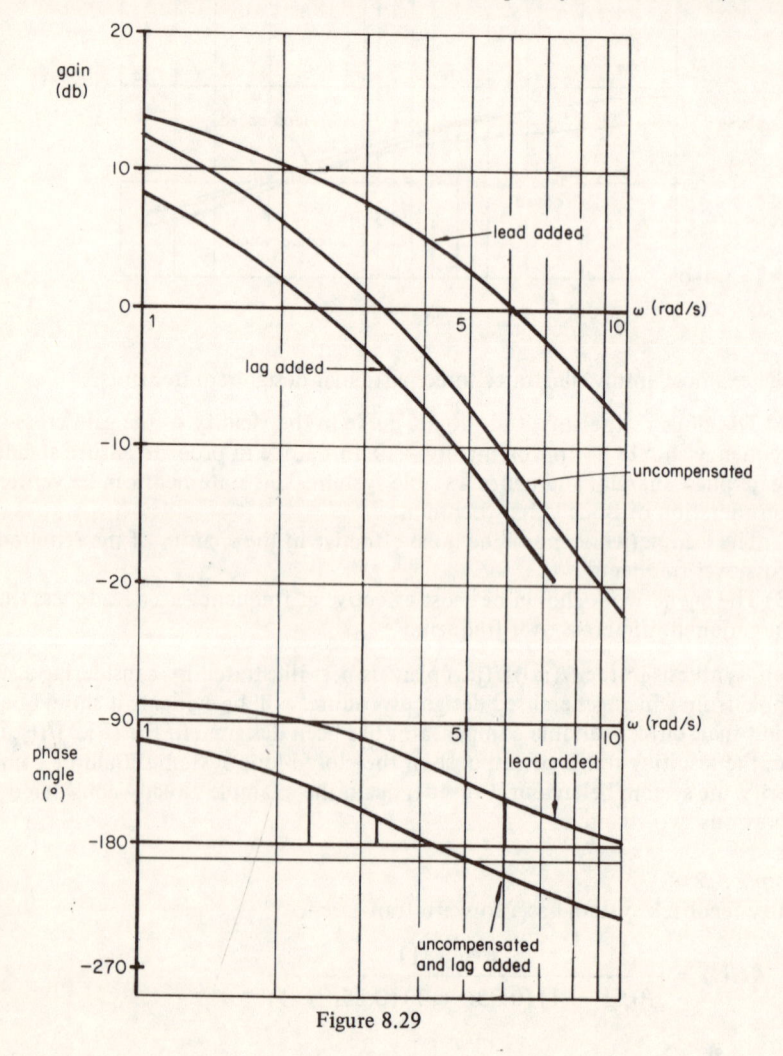

Figure 8.29

From figure 8.29 the appropriate gain modification is 7.37 db; if this is to be provided by the lead network at a frequency of 5 rad/s and is simultaneously to coincide with the maximum phase angle, α must be chosen such that

$$\frac{1}{\sqrt{\alpha}} = 7.37 \text{ db}$$

or

$$\alpha = 0.18$$

and

$$\phi_{max} = \sin^{-1} \frac{1 - \alpha}{1 + \alpha} = 44°$$

The phase at 5 rad/s is $-190°$, and therefore the new phase margin will be $-10 + 44 = 34°$, which is too low.

Choose the gain cross-over frequency to be at 6 rad/s instead; this will require a gain modification of 11 db, which will give $\alpha = 0.0782$ and $\phi_{max} = 58°$. The phase at 6 rad/s is $-200°$ and therefore the new phase margin will be $-20° + 58° = 38°$, which is acceptable. Therefore

$$T = \frac{1}{\omega\sqrt{\alpha}} = \frac{1}{6\sqrt{0.0782}} = 0.6 \text{ s}$$

and

$$\alpha T = 0.047 \text{ s}$$

Thus the transfer function of the required lead network is

$$\frac{0.6s + 1}{0.047s + 1}$$

and the resulting lead-compensated frequency response is shown in figure 8.29. It is seen that the system now exhibits a phase margin of 38° and a gain margin of 10 db while the bandwidth has been increased from 3.9 rad/s to 7.6 rad/s.

The only comment needed is that the value of $\alpha (= 0.0782)$ of the lead network is a little small and usually a phase lead of 58° should be achieved by cascading two lead networks.

Thus the steps in the design procedure in the $G(j\omega)H(j\omega)$ plane when adding a lead network are as follows.

(1) Determine the gain K of the uncompensated system to satisfy the steady-state requirements and draw the resulting frequency response using that value of K.

(2) Estimate the new position of the gain cross-over frequency and, from the required gain modification, calculate α and the resulting ϕ_{max}.

(3) Determine the new phase margin using the value of ϕ_{max} and, if acceptable, determine the value of T and αT and the transfer function of the lead network. If not, repeat step (2).

(4) Draw the lead-compensated frequency response, check the value of the phase margin and also the gain margin.

(5) If the value of $\alpha \ll 0.1$ the use of two lead networks in cascade should be considered or the possibility of employing a lag network in place of the lead network.
(6) Determine the resulting transient response.

A similar example demonstrates the design of a lag-network compensator.

Example 8.6
Design a lag-network compensator to achieve the performance requirements of example 8.5 for the system of that example.

To satisfy the velocity error constant, the gain should be chosen as before and the resulting gain-compensated frequency response drawn. The frequency at which the phase angle is $-180°$ plus the required phase margin—that is, $-180 + 40 = -140°$—is 2.4 rad/s and this is chosen to be the new gain cross-over frequency. The gain at this frequency is 4.4 db and, since the time constants of the lag network will be chosen not to interfere with the system transient response, choose $1/\beta$, the gain of the network at infinite frequency, such that

$$\frac{1}{\beta} = -4.4 \text{ db} = 0.6026$$

that is

$$\beta = 1.66$$

Now choose the break point of the lag network at $1/T$ at least a decade lower than the new gain cross-over frequency, that is

$$\frac{1}{T} = 0.1$$

Therefore

$$1/\beta T = 0.06$$

Thus the transfer function of the lag network is

$$\frac{10s + 1}{16.6s + 1}$$

and the resulting frequency response is shown in figure 8.29; it is seen that the compensated system exhibits a phase margin of 40°, a gain margin of 8.5 db and a bandwidth of 3 rad/s. This reduced bandwidth will result in a slower transient response.

Thus the steps in the design procedure in the $G(j\omega)H(j\omega)$ plane when adding a lag network are as follows.

(1) Determine the gain K of the uncompensated system to satisfy the steady-state requirements and draw the resulting frequency response using that value of K.
(2) Determine the frequency at which the phase angle is $-180°$ plus the required phase margin and select this frequency as the new gain cross-over frequency.
(3) Select β such that the gain of the lag network at infinite frequency provides the required gain modification.

Control-system Synthesis

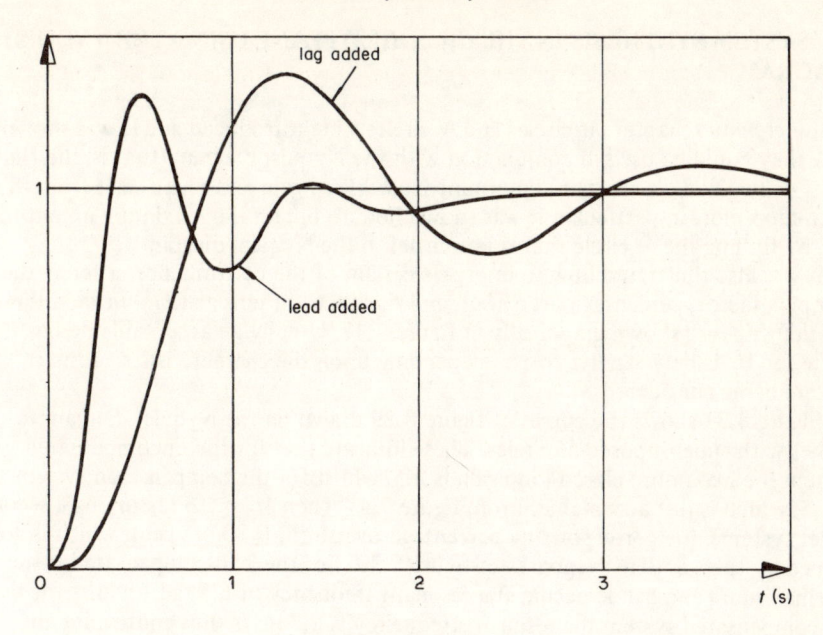

Figure 8.30

(4) Select the breakpoint at $1/T$ to be at least a decade lower than the new gain cross-over frequency. Calculate $1/\beta T$.
(5) Draw the lag-compensated frequency response and check the value of the phase margin and also the gain margin.
(6) Determine the resulting transient response.

It is constructive to compare the transient responses of the two systems resulting from examples 8.5 and 8.6; they are shown in figure 8.30. Although both systems have essentially the same gain and phase margins and hence similar stability characteristics, the transient response of the lead-compensated system is much faster than the system with the lag network. This is because the lead-compensated system has a larger bandwidth. This difference in system behaviour highlights the need to determine the resulting transient response.

Table 8.2 shows the closed-loop poles of the two systems.

Table 8.2

System with Lead	System with Lag
−0.85	−0.1
−1.71	−0.75 ± j2.86
−1.95 ± j6.93	−0.86
−23.8	−6.59

8.5 SYSTEM SYNTHESIS IN THE $G(j\omega)H(j\omega)$ PLANE USING THE NYQUIST DIAGRAM

In the previous chapter, M circles and N circles were introduced and it was shown how they could be used in conjunction with the Nyquist diagram (that is, the Hall chart or the Nichols chart) to determine closed-loop gain and phase shift from the open-loop plot. In particular, it was shown how to obtain the maximum amplitude M_p, by finding the M circle that is tangential to the Nyquist diagram.

It was also illustrated how to interpret certain of the performance criteria—the damping factor, percentage overshoot, and rise time—in terms of M_p. In fact, these relationships are shown graphically in figure 7.47. Usually an acceptable design value for M_p is between 1.1 to 1.5, depending upon the characteristics of the system being considered.

Figure 8.31 shows the curves of figure 8.29 drawn on the Nyquist diagram together with superimposed M circles. These indicate that for the uncompensated system the maximum closed-loop gain is 3.5, whilst for the compensated systems it is 1.5, which is just acceptable. From figure 7.47 when $M_p = 1.5$ (assuming a second-order system), the corresponding percentage overshoot is 30 per cent, which is confirmed by the transient responses of figure 8.30. For the lead-compensated system the maximum amplitude occurs at a resonant frequency of 6.5 rad/s whilst for the lag-compensated system the resonant frequency is 2.7 rad/s thus confirming the shorter rise time of the former.

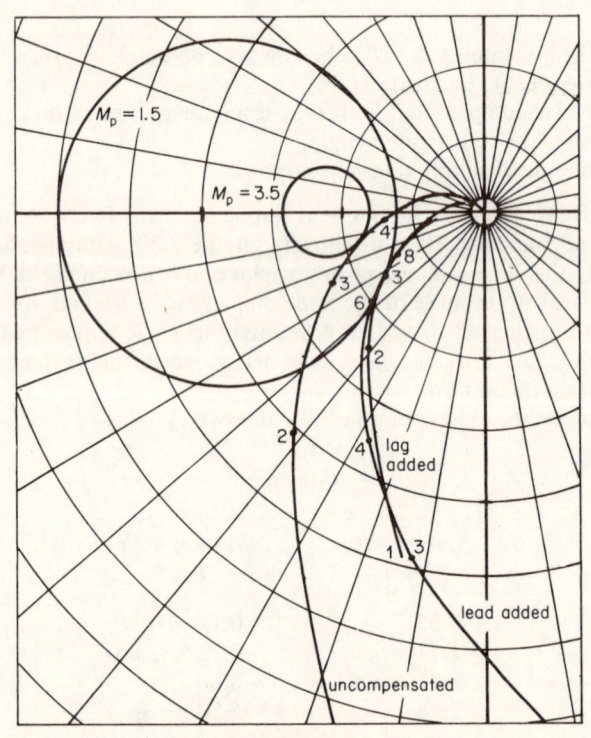

Figure 8.31

Control-system Synthesis

System synthesis using the Nyquist diagram to satisfy performance criteria in terms of M circles is illustrated by two examples; the first is concerned with the design of a lead network and the second with a lag network.

Example 8.7

Design a lead compensator for the unity-feedback control system with forward transference

$$K \frac{0.5s + 1}{s^2 (0.1667s + 1)(0.1s + 1)}$$

such that the acceleration error constant K_a is 3 and the maximum magnification M_p is less than or equal to 1.45.

The acceleration error constant K_a is given by

$$K_a = \lim_{s \to 0} [s^2 KG(s)]$$

that is

$$K_a = K$$

Therefore to satisfy the steady-state requirements the gain K must be chosen equal to 3.

Figure 8.32 shows the Nyquist diagram for the gain-compensated system with gain equal to 3, together with a superimposed M circle that indicates a maximum magnification of 4. Thus this system does not satisfy the transient-performance requirement and further compensation must be designed.

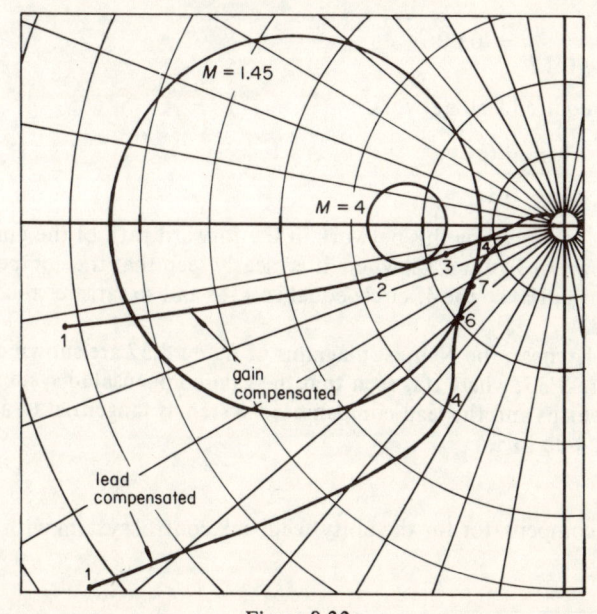

Figure 8.32

To design the required lead network first draw the M circle equal to 1.45; then if the lead-compensated system is to satisfy the transient-performance requirement, the corresponding Nyquist diagram must lie either touching this circle or outside it. From previous work in this chapter it is known that the inclusion of a lead network into a system increases the gain at high frequencies whilst reducing the phase lag within the system.

The method of design is essentially a trial-and-error method, although it is only necessary to consider the effect of the lead network over a small range of frequencies —in this particular example, between 4 and 8 rad/s. Noting that the phase angle at $\omega = 4$ rad/s should be advanced from $-172°$ to $-135°$, that is, by $37°$, the network chosen should advance the phase by $45°$ at a frequency of 4 rad/s so as to ensure that the Nyquist diagram for frequencies higher than 4 rad/s will not lie inside the design M circle.

The transfer function of a lead network is

$$G_c(s) = \frac{1 + Ts}{1 + \alpha Ts}$$

and since $\phi_{max} = 45°$

$$\frac{1 - \alpha}{1 + \alpha} = \sin 45°$$

and hence

$$\alpha = 0.17$$

This maximum phase advance should occur at the frequency $\omega = 4$ rad/s. Therefore

$$T = \frac{1}{4\sqrt{0.17}} = 0.606$$

Thus the required lead network is

$$G_c(s) = \frac{1 + 0.6s}{1 + 0.1s}$$

The effect of cascading this network in the forward path of the unity-feedback system is shown in figure 8.32, where it is clearly seen that the compensated Nyquist diagram is tangential to the M circle equal to 1.45 and so satisfies the performance requirements.

For completeness, the Nyquist diagrams of figure 8.32 are shown on the Nichols chart of figure 8.33, where it is seen that the gain-compensated system is tangential to the 12-db locus and the lead-compensated system is tangential to a locus between the 3-db and 4-db loci.

Example 8.8
Design a lag compensator for the unity-feedback control system with forward transference

$$\frac{K}{s(s + 1)(0.5s + 1)}$$

Control-system Synthesis

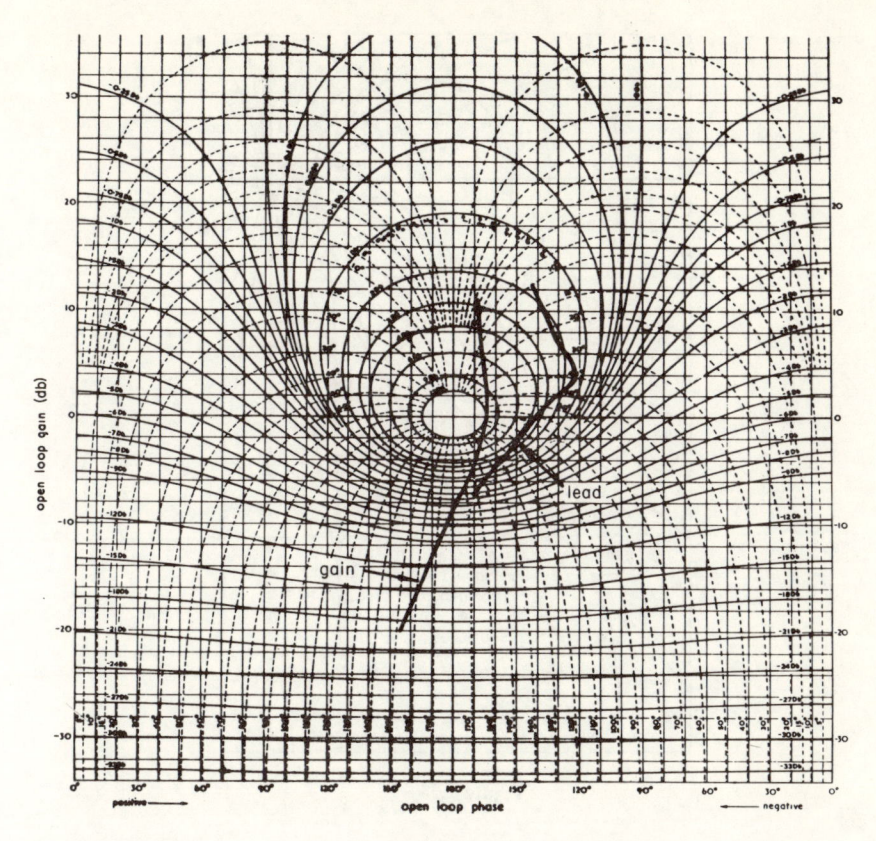

Figure 8.33

such that the velocity error constant, K_v, is 4 and the maximum magnification M_p is 1.4.

The velocity error constant is given by

$$K_v = \lim_{s \to 0} [sKG(s)]$$

that is

$$K_v = K$$

Therefore to satisfy the steady-state requirements, the gain K must be chosen equal to 4.

First draw the Nyquist diagram of the system for $K = 4$ and superimpose the M circle for $M = 1.4$. It is seen from figure 8.34 and by applying the Nyquist criterion that the system is unstable and the Nyquist diagram lies within the chosen M circle. It is now necessary to estimate the reduction in gain required for the Nyquist diagram to be tangential to $M = 1.4$. There are many ways to achieve this: construct a line from the origin tangential to the M circle (point A) until it meets the Nyquist diagram (point B). If the gain is reduced by the factor OB/OA the modified Nyquist diagram would be approximately tangential to the M circle but bearing in mind that

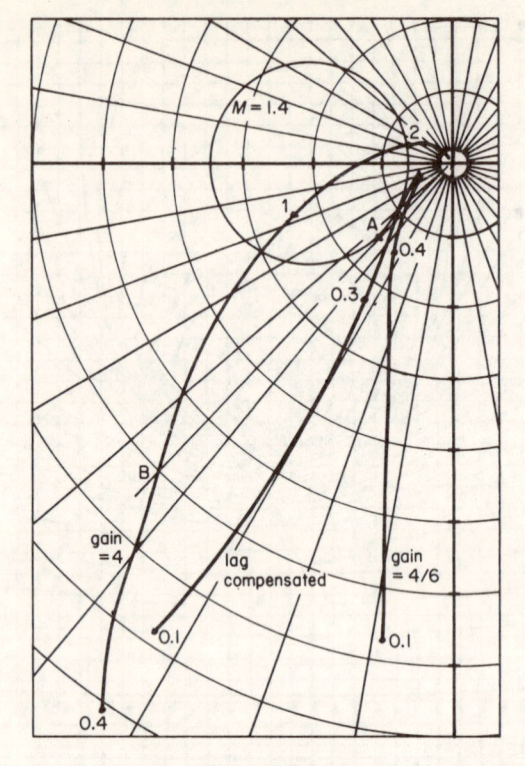

Figure 8.34

the required lag network will increase the phase lag, it is advisable to reduce the gain a little further.

From the diagram, OB/OA = 5 and so, as a first trial, reduce the gain by a factor of 6 and re-draw the Nyquist diagram with the new value of the gain, $K = 4/6$. This operation is relatively straightforward since the phase angle is unaffected and each gain value has only to be divided by 6. The modified diagram is shown in figure 8.34.

Now introduce into the system a lag network that is only effective in the low-frequency range, that is, a decade below the frequencies within the vicinity of the $M = 1.4$ circle. The low-frequency gain of this network must be equal to 6 in order to restore the velocity error constant and, since it will be designed so as not to be effective in the higher-frequency range, the transient-performance requirement will also be satisfied.

The lag-network transfer function is

$$G_c(s) = \beta \frac{(1 + Ts)}{(1 + \beta Ts)}$$

where $\beta = 6$ and $1/T$ is chosen to be 0.06 (that is, a decade below the frequency

Control-system Synthesis

$\omega = 0.6$ rad/s). Therefore $T = 16.67$ and $\beta T = 100$. Thus a satisfactory network should be

$$G_c(s) = 6\frac{(1 + 16.67s)}{(1 + 100s)}$$

and the compensated Nyquist diagram is given in figure 8.34.

The resulting transient response is shown in figure 8.35a, where it is seen that the overshoot is around 30 per cent, which corresponds to a maximum magnification of 1.4. The fairly long rise time corresponds to the narrow bandwidth of the lag-compensated system.

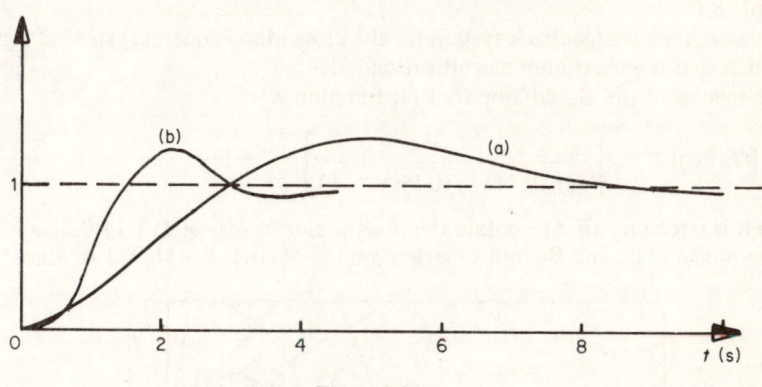

Figure 8.35

8.6 SYSTEM SYNTHESIS IN THE $G(j\omega)H(j\omega)$ PLANE USING THE INVERSE NYQUIST DIAGRAM

The inverse Nyquist diagram was illustrated in the previous chapter, where it was found to be particularly useful when considering the behaviour of closed-loop systems that contain subsidiary feedback loops. For example, the closed-loop transfer function of the system of figure 8.36 is

$$G'(j\omega) = \frac{KG(j\omega)}{1 + KG(j\omega)H(j\omega)} \qquad (8.6)$$

and therefore the inverse transfer function is

$$[G'(j\omega)]^{-1} = [KG(j\omega)]^{-1} + H(j\omega) \qquad (8.7)$$

$[G'(j\omega)]^{-1}$ may be obtained simply by the vector addition of the polar plots of $[KG(j\omega)]^{-1}$ and $H(j\omega)$, as shown in figure 7.48.

System synthesis using the inverse Nyquist diagram is illustrated in the next example.

204   Introduction to Control Theory

Figure 8.36

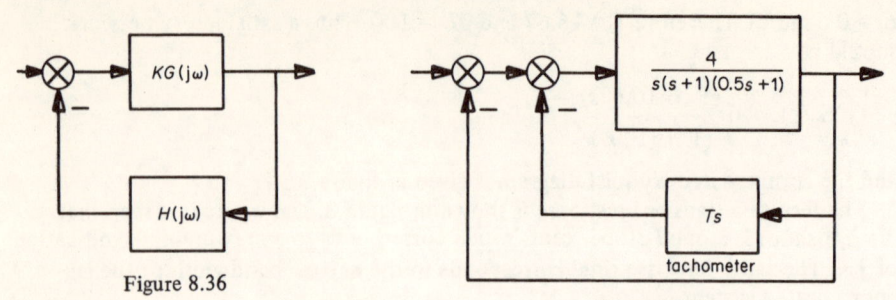

Figure 8.37

Example 8.9
Design a tachometer feedback system for the closed-loop control system of figure 8.37 such that the maximum magnification is 1.4.

The inverse of the closed-loop transfer function is

$$[G'(j\omega)]^{-1} = \left[\frac{4}{j\omega(j\omega + 1)(0.5j\omega + 1)}\right]^{-1} + 1 + Tj\omega$$

and so it is necessary first to obtain the inverse plot of $4/[j\omega(j\omega + 1)(0.5j\omega + 1)]$ and the plot of $Tj\omega$ and then to superimpose the M circle for $M_p = 1.4$. Since $Tj\omega$

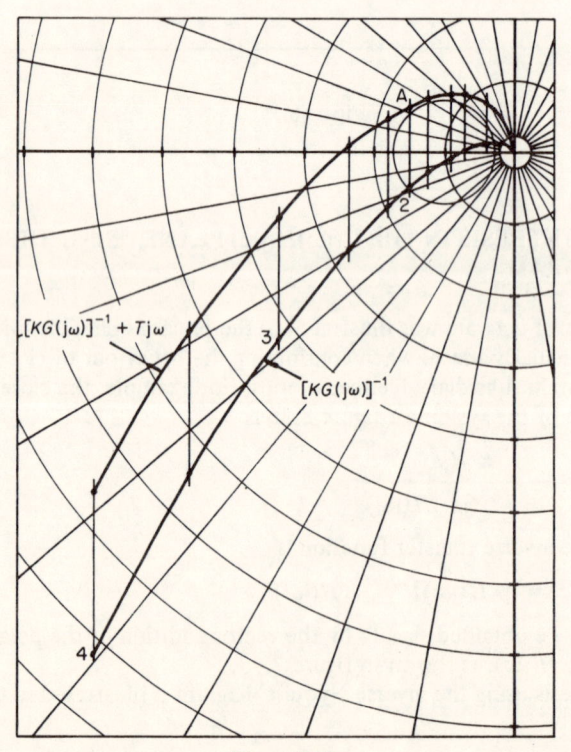

Figure 8.38

is to be plotted rather than $1 + Tj\omega$ the M circle for $M = 1.4$ in the inverse plane will be simply a circle with centre $(-1, 0)$ and radius $1/1.4 = 0.714$ and the critical point will be $(-1, 0)$. (If $1 + Tj\omega$ is plotted, the circle would have to be centred at the origin and the critical point would also be the origin. This important distinction must always be borne in mind when working in the inverse plane.)

Figure 8.38 shows the inverse plot of $KG(j\omega)$ and the superimposed M circle. The plot of $Tj\omega$ is a straight line along the positive imaginary axis and so, to obtain the plot of $[KG(j\omega)]^{-1} + Tj\omega$ all that is necessary is to add to each point on the $[KG(j\omega)]^{-1}$ plot a vector parallel to the imaginary axis and proportional in length to the frequency value at the point. To obtain a final plot that is tangential to the M circle the $[KG(j\omega)]^{-1}$ plot must be shifted upwards; for example, at the frequency $\omega = 2$ rad/s construct a vector parallel to the imaginary axis of length 1.1, at which point the vector will intersect the M circle (point A).

Thus at $\omega = 2$, the length of the additional vector is $\omega T = 1.1$. Therefore $T = 1.1/2 = 0.55$. Now plot $[KG(j\omega)]^{-1} + 0.55j\omega$ by vector addition; it is seen in figure 8.38 that the plot is tangential to the M circle, thus satisfying the transient-performance requirement. It is an easy matter to show also that the system satisfies the velocity-error requirement of the previous chapter.

The resulting transient response is shown in figure 8.35b.

8.7 FEEDFORWARD CONTROL

Consider a closed-loop system under the influence of a disturbance that is assumed to be measurable (see figure 8.39). Following a change in the disturbance, the output $c(s)$ will deviate from its desired value, and an error signal will be produced. If the plant transfer functions $G(s)$ and $G_1(s)$ contain large time lags, it will take an appreciable time before the feedback action corrects the effects of the change in the disturbance. This occurs especially in large process plant such as the boiler-turbine unit discussed in chapter 3. For example, a change in turbine speed will affect the steam flow (disturbance) through the turbine governor valve. This will cause a change in steam pressure, thus producing an error signal which will affect the firing rate. Because of the large thermal inertia of the boiler, the steam pressure

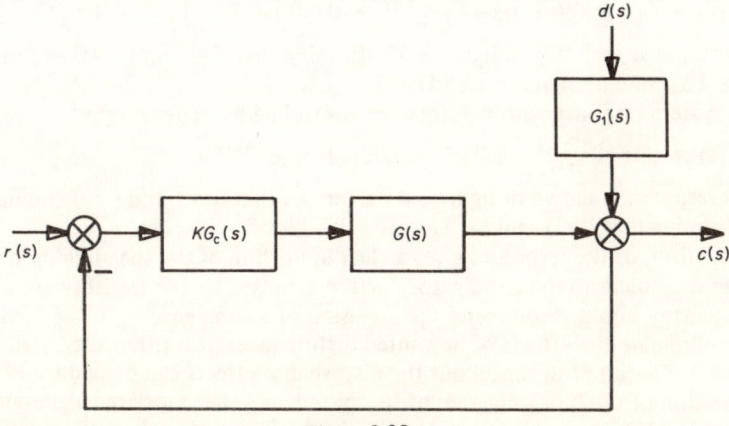

Figure 8.39

will be off its design value for an appreciable length of time.

A method used extensively in the process industries to speed up the system response is *feedforward control*. Essentially the steam flow disturbance is fed forward through a proportionality factor and the resulting signal is added to the output of the controller. Thus the effect of the feedforward control allows corrective action to be taken as soon as the disturbance changes, without the need to wait until the output changes.

It should be noted that feedforward control is open-loop control and should never be used except in conjunction with an associated feedback loop.

Figure 8.40 shows a simplified arrangement of a closed-loop control system with added feedforward control and, to show its effect, the system behaviour will be analysed with and without the addition.

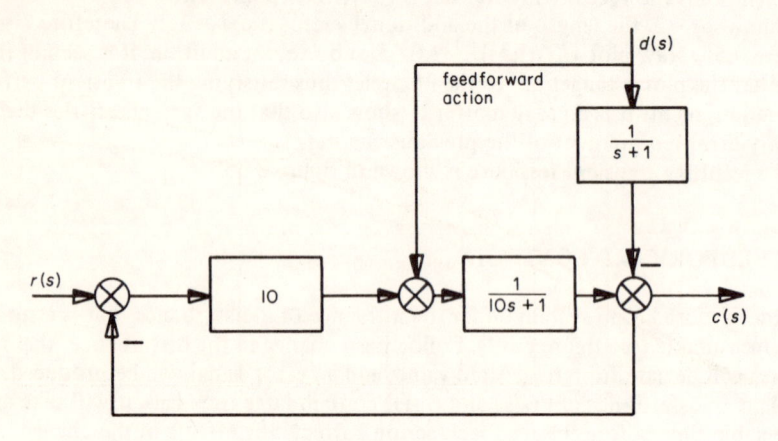

Figure 8.40

The system response without the additional feedforward control, assuming that both $r(s)$ and $d(s)$ are step changes, is

$$c(t) = (-0.0909 + 0.0433e^{-1.1t} + 0.0476e^{-0.05t})d + 0.909(1 - e^{-1.1t})r \quad (8.8)$$

This response is shown in figure 8.41 assuming that $d = 1$ and $r = 0$ and in figure 8.42 assuming that $d = 1$ and $r = 1$.

The system response with feedforward control added is given by

$$c(t) = 9.0(e^{-1.1t} - e^{-t})d + 0.909(1 - e^{-1.1t})r$$

This response is shown in figure 8.41 assuming that $d = 1$ and $r = 0$ and in figure 8.42 assuming that $d = 1$ and $r = 1$.

Comparison of the responses gives a clear indication of the speeding up of the response due solely to the anticipatory action afforded by the feedforward control.

While many books recommend the inclusion of a compensator in the feedforward path to eliminate the effects of unwanted disturbances, it is often suggested that the compensator required to cancel out the disturbance effects can be deduced from a consideration of the block diagram of the system and the associated algebraic relationships. However, the expression for the feedforward-path compensator

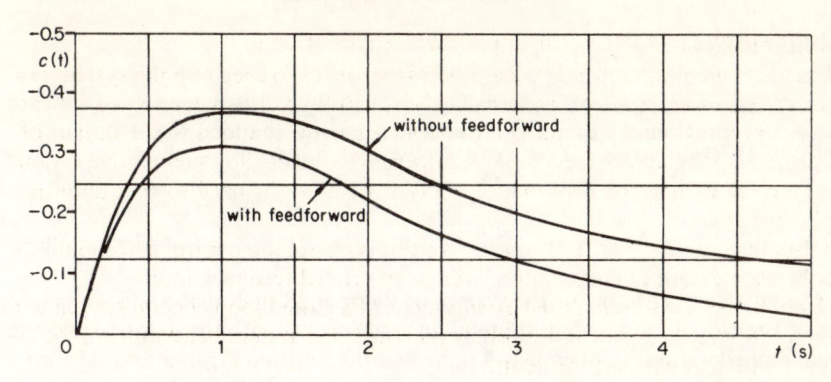

Figure 8.41

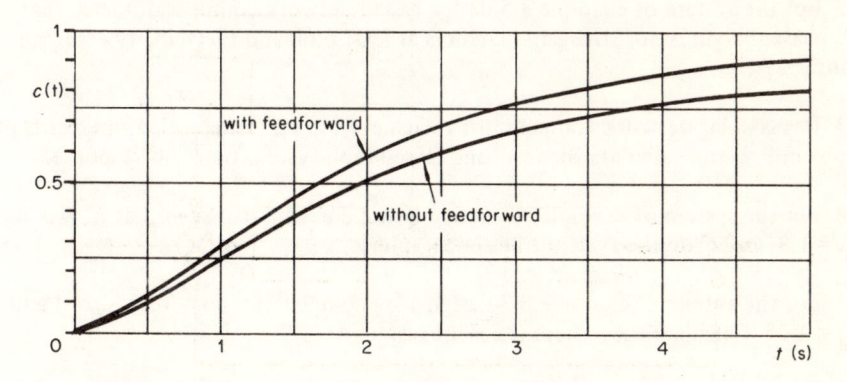

Figure 8.42

obtained in this way is usually fairly complicated and in many cases unrealisable, so the procedure is hardly ever used in practice, since a simpler compensator will usually suffice.

8.8 CONCLUDING REMARKS

Many methods of system synthesis have been presented in this chapter and the choice of a particular method must be based on the form of the performance requirement. There is no 'best' method of design as such, and usually in practice a designer becomes extremely adept at using one particular method, for example, the root-locus method, and applies it in most cases. The root-locus method does hold one slight advantage in that the closed-loop poles are more easily obtainable and hence the calculation of the system transient response is made easier.

With the wide usage of graphics terminals in both universities and industry all methods are easily programmed into the computer and, because of the speed of both computer and graphics terminal, the trial-and-error aspects of all the synthesis methods present no difficulties whatsoever.

REFERENCES

Kuo, B. C., *Automatic Control Systems* (Prentice-Hall, Englewood Cliffs, N. J., 1967).
Ogata, K., *Modern Control Engineering* (Prentice-Hall, Englewood Cliffs, N. J., 1970).
Shinners, S. M., *Control System Design* (Wiley, New York, 1964).

PROBLEMS

8.1 Design a compensator for the system of figure 8.23 such that the dominant poles of the closed-loop system subtend an angle of 45° with the negative real axis, the undamped natural frequency is 4 rad/s and the position error constant is not less than 19.

8.2 For the system of example 8.5 design a lead-network compensator such that the phase margin is 30°, the gain margin is at least 6 db and the velocity error constant is 3.5.

8.3 Design a lag-network compensator to achieve the performance requirements of problem 8.2, and calculate the resulting closed-loop-system transient response.

8.4 For the system of example 8.7 design a lead compensator such that $K_a = 3$ and $M_p = 1.3$. Calculate the system transient response.

8.5 For the system of example 8.8 design a lag compensator such that $K_v = 4$ and $M_p = 1.25$. Calculate the system transient response.

8.6 For the system of figure 8.36, $KG(s) = 70\,000/[s(as + b)]$ and $H(s) = 1 + 0.00457s$. Calculate a and b such that the system has a maximum magnification of 1.2 at a resonant frequency of 100 rad/s.

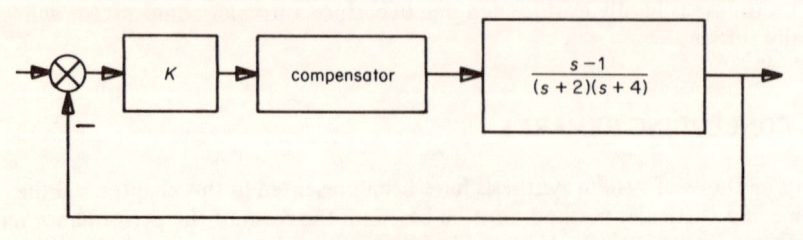

Figure 8.43

8.7 Choose a lead compensator for the system of figure 8.43 such that the system is stable for all K satisfying $0 < K < K_1$ for some $K_1 > 8$. Sketch the resulting root-locus diagram and calculate the maximum gain for stability.

8.8 For the system of figure 8.36 $KG(s) = 1250/[s(s + 12.5)(s + 8.33)]$ and $H(s) = 1 + Ts$; determine the value of T such that $M_p = 1.3$.

Control-system Synthesis

8.9 The forward transference of a unity-feedback system is

$$KG(s) = \frac{10\,000K}{s(s + 10)(s + 1000)}$$

Design a compensator such that $K_v = 1000$ and the phase margin is $45°$.

8.10 The forward transference of a unity-feedback system is

$$KG(s) = \frac{100K}{s(s + 5 + j8.66)(s + 5 - j8.66)}$$

Design a compensator such that $M_p = 1.2$ and $K_v = 6$.

9 Bridging the Gap

Many systems that are required to be controlled—for example, the boiler-turbine unit discussed in earlier chapters—consist of several inputs and several outputs, and are known as multivariable systems, whereas the systems that have been analysed so far throughout this book are exclusively single-input single-output (SISO) systems. Thus it should be asked at this stage whether the SISO techniques are applicable to multivariable systems. Unfortunately, the techniques are not directly applicable, because of the presence of interaction between the variables. Figure 9.1 represents a

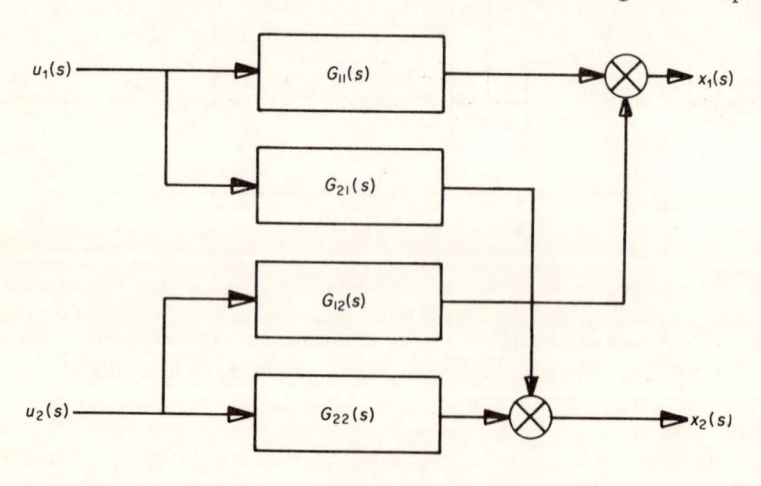

Figure 9.1

multivariable system with two inputs $u_1(s)$ and $u_2(s)$ and two outputs $x_1(s)$ and $x_2(s)$, where

$$\left. \begin{array}{l} x_1(s) = G_{11}(s)u_1(s) + G_{12}(s)u_2(s) \\ x_2(s) = G_{21}(s)u_1(s) + G_{22}(s)u_2(s) \end{array} \right\} \quad (9.1)$$

The above-mentioned interaction is contributed by the transfer functions $G_{12}(s)$ and $G_{21}(s)$, which connect output 2 to input 1 and output 1 to input 2, respectively

Bridging the Gap

If the transfer functions $G_{12}(s)$ and $G_{21}(s)$ were both zero, the multivariable system would reduce to two independent SISO systems, that is

$$\left. \begin{array}{c} x_1(s) = G_{11}(s)u_1(s) \\ x_2(s) = G_{22}(s)u_2(s) \end{array} \right\} \quad (9.2)$$

which can be analysed using the classical techniques of the previous chapters.

The multivariable system of figure 9.1 is shown again in figure 9.2 with added feedback control. If the controllers are designed assuming that the cross-coupling transfer functions $G_{12}(s)$ and $G_{21}(s)$ are not present—that is as though the system

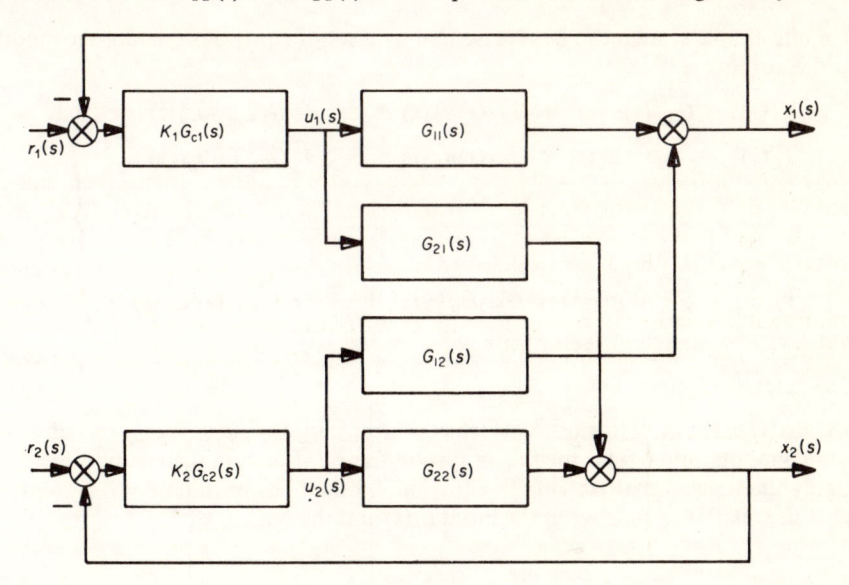

Figure 9.2

consisted of two independent SISO systems—this would usually result in a multivariable system that was less stable than either of the two SISO systems. Thus the presence of interaction within the multivariable system tends to reduce the stability of the system, causes deterioration in system response, and can cause instability.

This final chapter is intended to introduce the reader to the design of control systems for multivariable systems and to give him an appreciation of the principles involved; essentially it 'bridges the gap' between classical control theory and a particular branch of present-day work known as the inverse-Nyquist-array technique, pioneered by Professor H. H. Rosenbrock of the Control-systems Centre at the University of Manchester Institute of Science and Technology, England. This technique is essentially a multivariable frequency-response technique, which depends upon an understanding of the interaction between variables and on the availability of computer graphics, that is, graphics terminals interfaced to digital computers.

No mention is made of optimal control theory since, in the author's opinion, it has limited practical application to the control of process systems. This particular

Introduction to Control Theory

theory is based on the minimisation of mathematical performance measures that are difficult to interpret from an engineering point of view, whereas the inverse-Nyquist-array techniques allow familiar engineering constraints and performance requirements to be satisfied.

Throughout the chapter, vector and matrix algebra is used where appropriate and is fully explained.

9.1 REPRESENTATION OF MULTIVARIABLE SYSTEMS

A multivariable system can be represented by a set of equations relating the inputs to the outputs

$$\left.\begin{array}{l} x_1(s) = G_{11}(s)u_1(s) + G_{12}(s)u_2(s) + \ldots + G_{1m}(s)u_m(s) \\ x_2(s) = G_{21}(s)u_1(s) + G_{22}(s)u_2(s) + \ldots + G_{2m}(s)u_m(s) \\ \quad \vdots \\ x_m(s) = G_{m1}(s)u_1(s) + G_{m2}(s)u_2(s) + \ldots + G_{mm}(s)u_m(s) \end{array}\right\} \quad (9.3)$$

which can be written in vector form as

$$x(s) = G(s)u(s) \quad (9.4)$$

where $x(s)$ is the m vector of transformed outputs; $u(s)$ is the m vector of transformed inputs; and $G(s)$ is the $m \times m$ system transfer-function matrix, or more simply the transfer matrix. Thus in equation 9.3 the transfer function $G_{ij}(s)$ represents the relationship between the input $u_j(s)$ and the output $x_i(s)$.

Example 9.1

Determine the inverse of the transfer-function matrix $G(s)$ where

$$G(s) = \begin{bmatrix} \dfrac{1}{(s+1)^2} & \dfrac{1}{(s+1)(s+2)} \\ \dfrac{1}{(s+1)(s+2)} & \dfrac{s+3}{(s+2)^2} \end{bmatrix}$$

For an $n \times n$ matrix A its inverse A^{-1} is defined by

$$A^{-1} = \frac{\text{adj } A}{\det A} \quad (\det A \neq 0)$$

where adj A is the transposed matrix of cofactors and det A is the determinant of A.

Before calculating the inverse, first obtain the common denominator of the elements of $G(s)$, which in this example is $(s+1)^2(s+2)^2$, and re-write $G(s)$ in terms of this common denominator

Bridging the Gap

$$G(s) = \frac{1}{(s+1)^2(s+2)^2}\begin{bmatrix} (s+2)^2 & (s+1)(s+2) \\ (s+1)(s+2) & (s+3)(s+1)^2 \end{bmatrix}$$

Therefore

$$G(s)^{-1} = (s+1)^2(s+2)^2 \frac{\text{adj } G(s)}{\det G(s)}$$

where

$$\text{adj } G(s) = \begin{bmatrix} (s+3)(s+1)^2 & -(s+1)(s+2) \\ -(s+1)(s+2) & (s+2)^2 \end{bmatrix}$$

and

$$\det G(s) = (s+2)^2(s+3)(s+1)^2 - (s+1)^2(s+2)^2$$

Therefore

$$G(s)^{-1} = \frac{(s+1)^2(s+2)^2}{(s+2)^2(s+3)(s+1)^2 - (s+1)^2(s+2)^2}$$

$$\begin{bmatrix} (s+3)(s+1)^2 & -(s+1)(s+2) \\ -(s+1)(s+2) & (s+2)^2 \end{bmatrix}$$

$$= \frac{1}{s+2}\begin{bmatrix} (s+3)(s+1)^2 & -(s+1)(s+2) \\ -(s+1)(s+2) & (s+2)^2 \end{bmatrix}$$

This should be checked, and so it is necessary to form $G(s)G(s)^{-1}$, as follows

$$G(s)G(s)^{-1} = \frac{1}{(s+1)^2(s+2)^3}$$

$$\begin{bmatrix} (s+3)(s+1)^2(s+2)^2 - (s+1)^2(s+2)^2 & 0 \\ 0 & -(s+1)^2(s+2)^2 + (s+2)^2(s+3)(s+1)^2 \end{bmatrix}$$

$$= \begin{bmatrix} 1 & 0 \\ 0 & 1 \end{bmatrix}$$

This is the unit matrix, and it thus follows that $G(s)^{-1}$ is correct.

A multivariable system with added feedback control is shown in figure 9.3 where $r(s)$ is the m vector of reference input transforms; $x(s)$ is the m vector of output

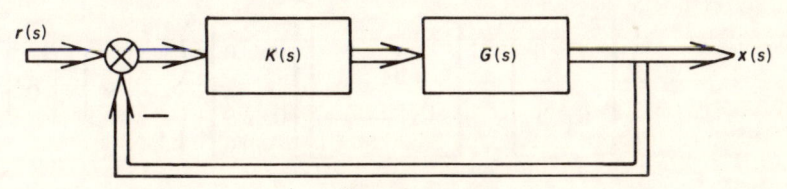

Figure 9.3

transforms; $u(s)$ is the m vector of input transforms; $K(s)$ is the $m \times m$ matrix of controller transfer functions; and $G(s)$ is the $m \times m$ matrix of transfer functions. $K(s)$ and $G(s)$ are rational functions in the complex variable s.

From figure 9.3

$$x(s) = G(s)u(s) \tag{9.4}$$

$$u(s) = K(s)[r(s) - x(s)] \tag{9.5}$$

and therefore

$$x(s) = G(s)K(s)[r(s) - x(s)] \tag{9.6}$$

that is

$$x(s) = [I_m + G(s)K(s)]^{-1} G(s)K(s)r(s) \tag{9.7}$$

where I_m is the $m \times m$ unit matrix and the closed-loop transfer-function matrix $G'(s)$ is

$$G'(s) = [I_m + G(s)K(s)]^{-1} G(s)K(s) \tag{9.8}$$

Since, in general, for two $n \times n$ matrices A and B

$$AB \neq BA \tag{9.9}$$

the order of $G(s)$ and $K(s)$ in equation 9.8 is most important, whereas for the SISO system the order is immaterial.

Example 9.2

Investigate the step response of the multivariable system of figure 9.2 when $K_1 = K_2 = 9$, $G_{c1}(s) = G_{c2}(s) = 1$, $G_{11}(s) = 1/(s+1)$ and $G_{22}(s) = 1/(0.5s+1)$ and when

(i) $G_{12}(s) = G_{21}(s) = 0$
(ii) $G_{12}(s) = 1$; $G_{21}(s) = -1$

(i) When $G_{12}(s) = G_{21}(s) = 0$, the multivariable system reduces to two independent SISO systems with closed-loop step responses

$$x_1(t) = 0.9(1 - e^{-10t})$$

$$x_2(t) = 0.9(1 - e^{-20t})$$

and these are shown in figures 9.4a and b, respectively.

(ii) When $G_{12}(s) = 1$ and $G_{21}(s) = -1$, equation 9.7 takes the form

$$\begin{bmatrix} x_1(s) \\ x_2(s) \end{bmatrix} = x(s)$$

$$= \left\{ \begin{bmatrix} 1 & 0 \\ 0 & 1 \end{bmatrix} + \begin{bmatrix} \dfrac{1}{s+1} & 1 \\ -1 & \dfrac{1}{0.5s+1} \end{bmatrix} \begin{bmatrix} 9 & 0 \\ 0 & 9 \end{bmatrix} \right\}^{-1} \begin{bmatrix} G \end{bmatrix} \begin{bmatrix} K \end{bmatrix} r(s)$$

The system response to step changes in the reference inputs can be obtained either by simplifying the above equation and then using inverse Laplace transforms; or by using an analog computer and patching up the block diagram of figure 9.2; or by replacing s by d/dt in the block diagram (since the initial conditions are zero), setting up the differential equations and solving either exactly or numerically. Usually the second or third alternatives are preferred and in fact the latter was used in this example.

The responses of the outputs $x_1(t)$ and $x_2(t)$ for $r_1(t) = 1, r_2(t) = 0$ and for $r_1(t) = 0, r_2(t) = 1$ are shown in figures 9.4a and b respectively. It is seen that, when the off-diagonal terms are present, the response of $x_1(t)$ to a step change in $r_1(t)$ is slower and slightly oscillatory and the variable $x_2(t)$ also experiences a change. Similarly the response of $x_2(t)$ to a step change in $r_2(t)$ is also slower and the variable $x_1(t)$ experiences a large negative response. The changes of $x_1(t)$ and $x_2(t)$ to changes in $r_2(t)$ and $r_1(t)$, respectively, are due to the off-diagonal terms in $G(s)$ and are crude measures of the amount of interaction present within the transfer function matrix $G(s)$.

The aim of the particular multivariable frequency-response technique known as the inverse-Nyquist-array technique is first to obtain a measure of this interaction, second to design a multivariable controller to reduce the interaction, and third to design single-loop controllers to achieve acceptable system responses. This is explained and illustrated in the next section.

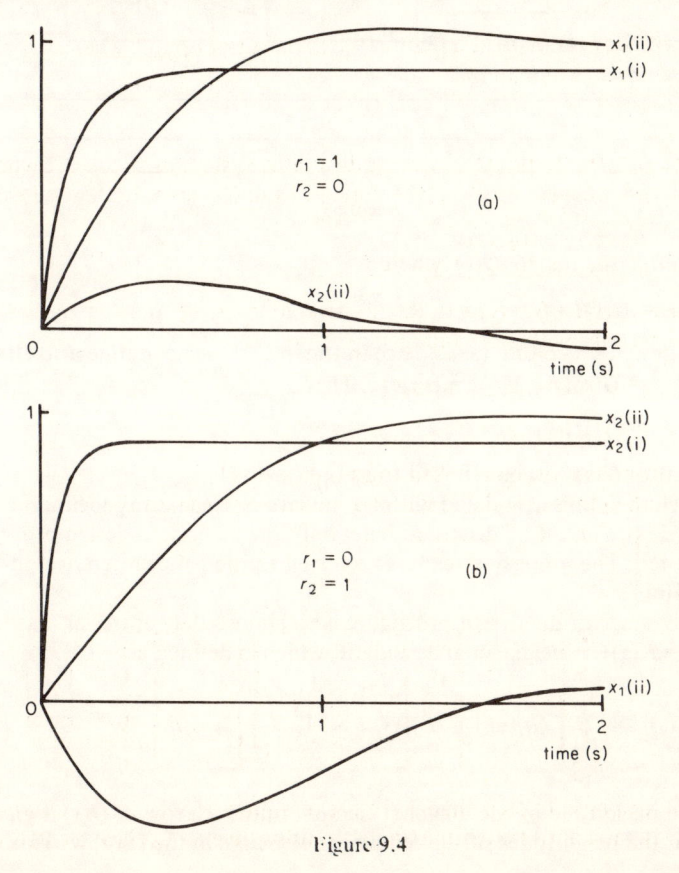

Figure 9.4

9.2 THE INVERSE-NYQUIST-ARRAY TECHNIQUE

From figures 9.5a and b, $K(s)$ is an $m \times m$ matrix of controller transfer functions and $K_i(s)$ is the $m \times m$ diagonal matrix of additional controller transfer functions and may be expressed as diag $[k_i(s)]$. The reason for defining two controller matrices will become evident later on. It is assumed that the open-loop responses arising from $G(s)$ are asymptotically stable and that both det $G(s)$ and det $K(s)$ have no zeros in the right-half-plane.

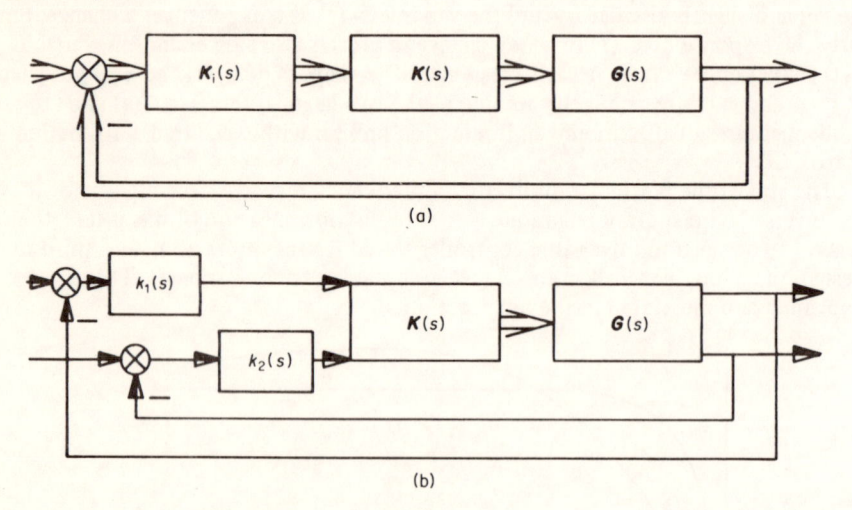

Figure 9.5

Now define the matrix $Q(s)$, where

$$Q(s) = G(s)K(s) \qquad (9.10)$$

Since the design procedure takes place in the inverse plane, define also $Q(s)^{-1} = \hat{Q}(s)$, $G(s)^{-1} = \hat{G}(s)$, $K(s)^{-1} = \hat{K}(s)$, etc. Then

$$\hat{Q}(s) = \hat{K}(s)\hat{G}(s) \qquad (9.11)$$

An element of $\hat{G}(s)$ will be referred to as $\hat{G}_{ij}(s)$.

Stability theorems associated with the inverse-Nyquist-array technique that are only sufficient will not be discussed here; only the associated design procedure will be considered. The interested reader is referred to the list of references at the end of the chapter.

The first stage in the design procedure is to choose the controller matrix $K(s)$ such that $\hat{Q}(s) = \hat{K}(s)\hat{G}(s)$ is diagonal dominant, which is defined as

$$|\hat{Q}_{ii}(s)| \geqslant \sum_{\substack{j=1 \\ i \neq j}}^{m} |\hat{Q}_{ij}(s)| = D_i(s), \; i = 1, 2, \ldots, m \qquad (9.12)$$

that is, the magnitude of the diagonal element in the ith row of $\hat{Q}(s)$ is greater than the sum of the magnitudes of the remaining elements in the ith row. This condition

Bridging the Gap

can be determined graphically by first drawing the inverse Nyquist diagrams for **each** of the elements of $\tilde{Q}(s)$ for $0 < \omega < \infty$, that is, the inverse Nyquist array, and then superimposing on the principal inverse Nyquist diagrams (that is, those corresponding to the diagonal elements in each row) circles centred at each particular frequency value on these diagrams with radii equal to the $D_i(\omega)$. If the band swept out by the circles does not contain the origin, $\tilde{Q}(s)$ is diagonal dominant. This graphical procedure is illustrated in figure 9.6. Further, if these bands intersect the negative real axis at points $-c_i$ (see for example figure 9.9), the stability of the closed-loop multivariable system is ensured in each loop for gains up to c_i.

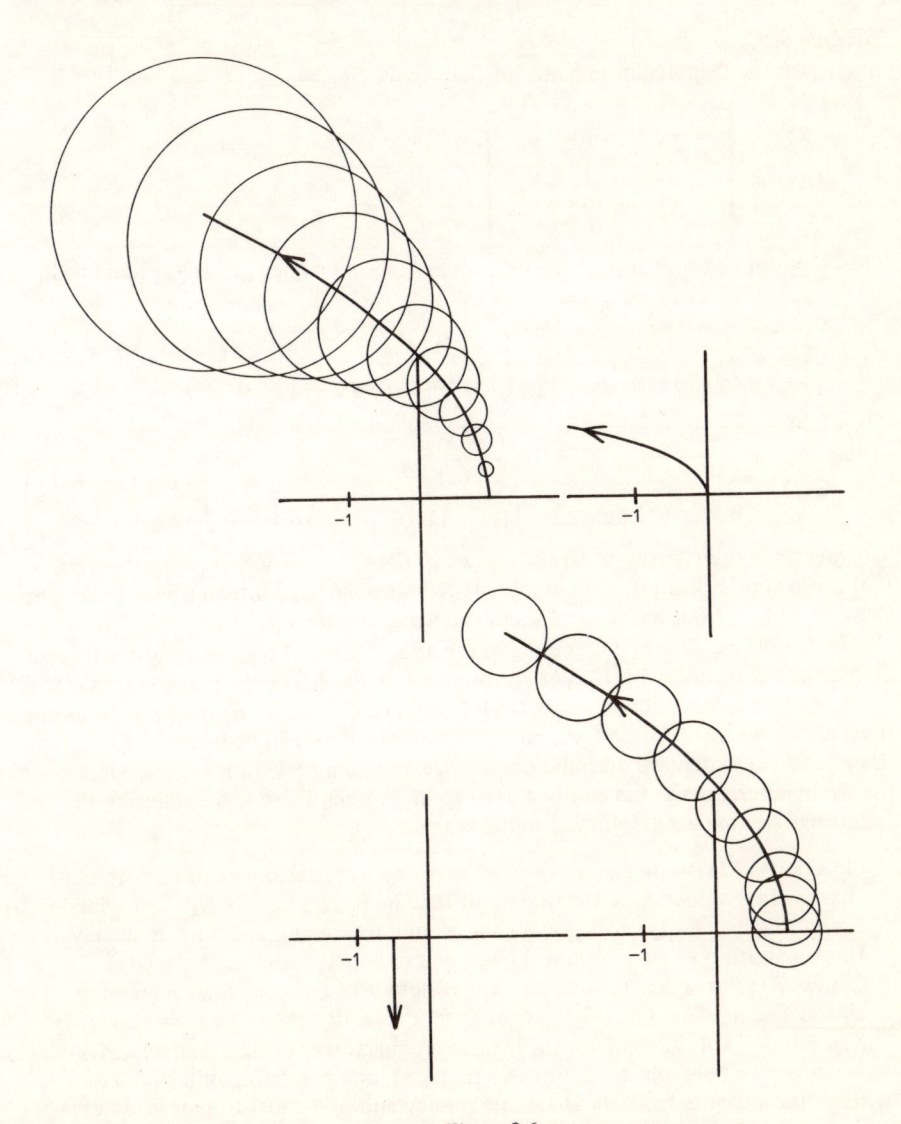

Figure 9.6

The band swept out by the circles is a measure of the degree of interaction present within the system and, although it is fairly tedious to produce figures such as figure 9.6 by hand, it is a very simple matter if access to a graphics terminal is available.

Once diagonal dominance has been established, the controller matrix $K_i(s) = \text{diag}[k_i(s)]$ can easily be designed by designing the individual controllers $k_i(s)$ using the single-loop theory of the previous chapters and considering the system to be controlled as being the appropriate diagonal element of $Q(s)$ and neglecting the effect of the off-diagonal terms.

Example 9.3
Investigate the diagonal dominance of the matrix $\hat{G}(s)$ where

$$G(s) = \begin{bmatrix} \dfrac{1}{s+1} & 1 \\ -1 & \dfrac{1}{0.5s+1} \end{bmatrix}$$

The common denominator of $G(s)$ is $(s+1)(0.5s+1)$ and so $G(s)$ can be written as

$$G(s) = \frac{1}{(s+1)(0.5s+1)} \begin{bmatrix} 0.5s+1 & (s+1)(0.5s+1) \\ -(s+1)(0.5s+1) & s+1 \end{bmatrix}$$

Therefore

$$\hat{G}(s) = \frac{1}{0.5s^2+1.5s+2} \begin{bmatrix} s+1 & -(s+1)(0.5s+1) \\ (s+1)(0.5s+1) & 0.5s+1 \end{bmatrix}$$

The inverse Nyquist array of $\hat{G}(s)$ is shown in figure 9.7, where it is clearly seen that the system is not diagonal dominant and the degree of interaction is large. This is confirmed by example 9.2 and the responses of figure 9.4.

The major problem in designing a multivariable control system using the inverse Nyquist array technique is that of choosing a suitable $K(s)$ such that $\hat{Q}(s) = \hat{K}(s)\hat{G}(s)$ is diagonal dominant. Obviously if $\hat{G}(s)$ is diagonal dominant, $\hat{K}(s) = I$; this condition must always be checked first. At present much effort is being concentrated on the design of $K(s)$ to achieve diagonal dominance and many techniques are contained in the list of references at the end of this chapter. Among these techniques are the following (assuming a graphics terminal is available).

(1) Choose $K(s)$ to be a constant matrix K and use trial and error to determine the constant values. One starting point may be to choose $\hat{K} = G(0)$; provided $G(0)$ is non-singular, $KG(s)$ is diagonal (that is, non-interacting) at zero frequency. Diagonalisation at other frequencies may also be tried, that is, $\hat{K} = |G(s)|_{s=j\omega_1}$. Choose $K(s)$ to be a matrix of lead or lag networks and use trial and error as above. The networks may be designed to be effective at low or high frequency depending upon the shape of the inverse Nyquist array or, alternatively, $K(s)$ could be chosen to be a summation of two matrices, one consisting of lag networks that diagonalise effectively at low frequency and one consisting of lead networks designed to be effective at high frequency.

Bridging the Gap

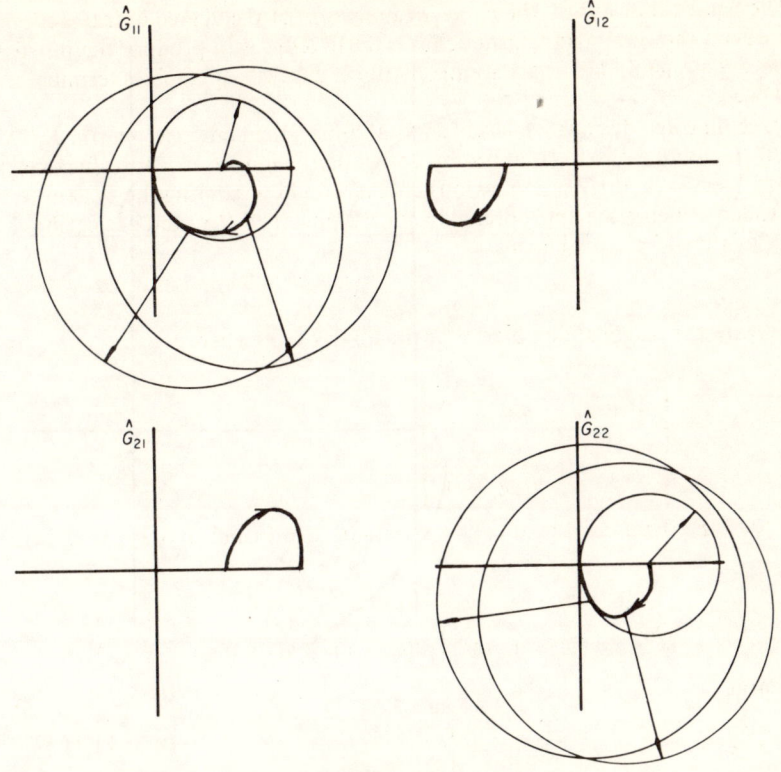

$\hat{G}_{11}$ $\hat{G}_{12}$

$\hat{G}_{21}$ $\hat{G}_{22}$

Figure 9.7

Consider the following example.

Example 9.4
Choose $\hat{K}(s)$ such that $\hat{K}(s)\hat{G}(s)$ is diagonal dominant where $G(s)$ is given by

$$G(s) = \begin{bmatrix} 0.048s & 0.001119s(s - 0.737) \\ -0.386s & 0.00192(s - 0.75)(s - 0.14) \end{bmatrix} \frac{1}{100s(s + 0.1)(s + 0.008)}$$

First compute the inverse transfer function $\hat{G}(s)$, which is

$$\hat{G}(s) = \frac{(10s + 1)(125s + 1)}{(1.35s - 1)(40.2s - 1)} \begin{bmatrix} 1.67(1.33s - 1)(7.19s - 1) & 6.85s(1.36s - 1) \\ 3208s & 398.9s \end{bmatrix}$$

and plot the inverse Nyquist array as shown in figure 9.8. Then superimpose circles of radius $|\hat{G}_{12}(s)|$ on the diagram of $\hat{G}_{11}(s)$ and circles of radius $|\hat{G}_{21}(s)|$ on the diagram of $\hat{G}_{22}(s)$. It is clearly seen that the first row is diagonal dominant, but the second row is not diagonal dominant since the circles enclose the origin.

Since the first row is dominant, choose $\hat{K}(s)$ such that

$$\hat{K}(s) = \frac{1}{\beta(s)} \begin{bmatrix} 1 & 0 \\ \alpha(s) & \beta(s) \end{bmatrix}$$

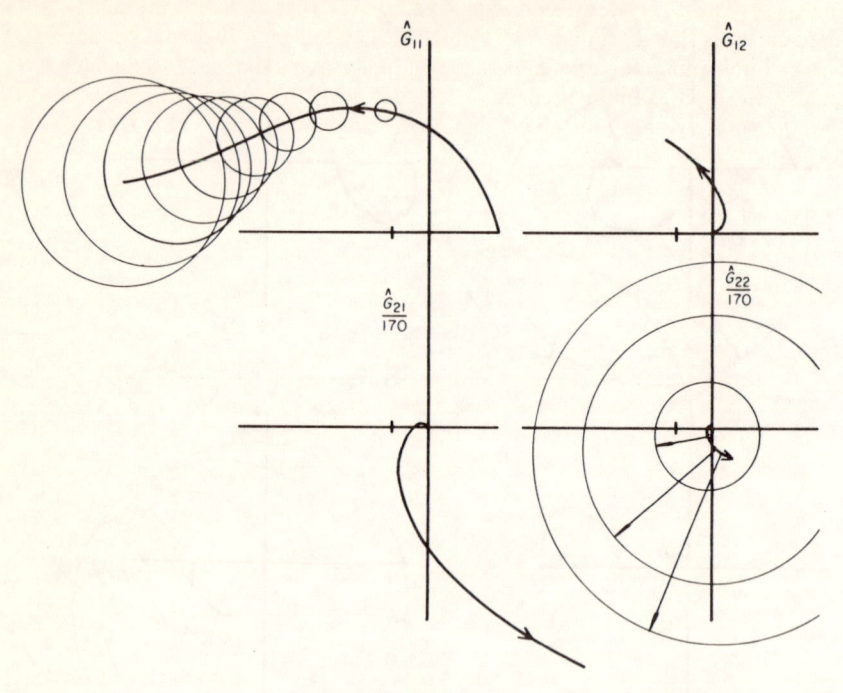

Figure 9.8

that is

$$K(s) = \begin{bmatrix} \beta(s) & 0 \\ -\alpha(s) & 1 \end{bmatrix}$$

which will obviously not affect the first row of $\hat{G}(s)$. Note that det $\hat{K}(s)$ is unity.

Let

$$\hat{G}(s) = E(s) \begin{bmatrix} A(s) & B(s) \\ C(s) & D(s) \end{bmatrix}$$

then

$$\hat{K}(s)\hat{G}(s) = \frac{E(s)}{\beta(s)} \begin{bmatrix} A(s) & B(s) \\ \alpha(s)A(s) + \beta(s)C(s) & \alpha(s)B(s) + \beta(s)D(s) \end{bmatrix}$$

Since $A(s) = 1.67(1.33s - 1)(7.19s - 1)$ and $C(s) = 3208s$ choose $\alpha(s) = -3208s$ and $\beta(s) = 1.67(1.33s - 1)(7.19s - 1) = 15.96(s^2 - 0.89s + 0.1)$. Since the frequency range of interest is less than 0.1 rad/s (this is known by plotting the inverse Nyquist array), the s^2 term can be safely neglected; therefore choose

$$\beta(s) = 15.96(-0.89s + 0.1) = -1.67(8.52s - 1)$$

Therefore

$$K(s) = \begin{bmatrix} -1.67(8.52s - 1) & 0 \\ 3208s & 1 \end{bmatrix}$$

Bridging the Gap

The inverse Nyquist array corresponding to $\hat{K}(s)\hat{G}(s)$, together with the superimposed cicles, is shown in figure 9.9, which clearly indicates diagonal dominance.

This example illustrates a possible method of achieving diagonal dominance.

Finally, the next example is chosen to illustrate the full design procedure, including the choice of $K(s)$ and the design of the single-loop controllers $K_i(s)$, to achieve satisfactory response.

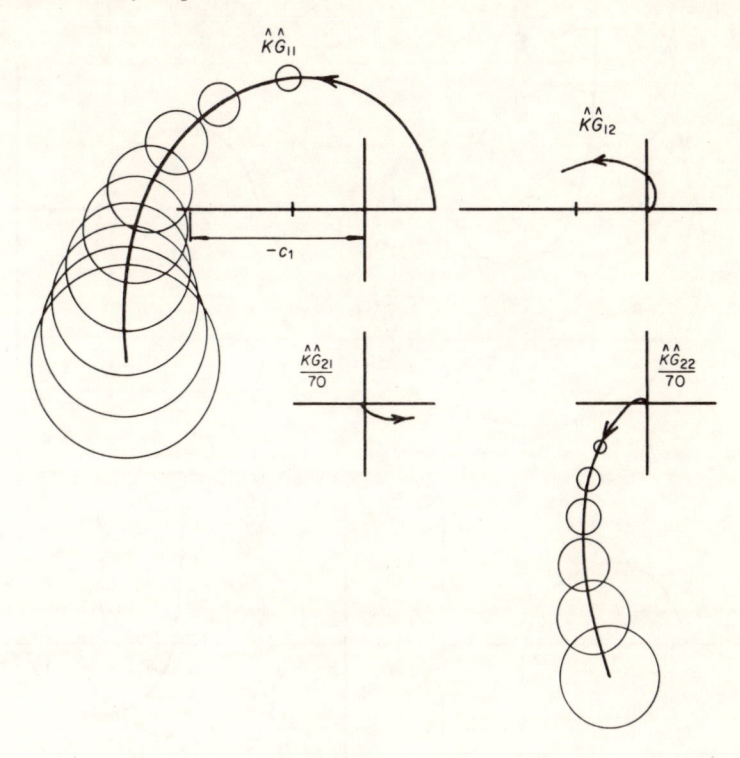

Figure 9.9

Example 9.5
Design a multivariable controller $K(s)$ to improve the diagonal dominance of the system with transfer-function matrix

$$G(s) = \begin{bmatrix} \dfrac{1}{s+1} & \dfrac{0.75}{2s+1} \\ \dfrac{0.75}{2s+1} & \dfrac{1}{0.5s+1} \end{bmatrix}$$

and choose single-loop controllers $k_1(s)$ and $k_2(s)$ such that the closed-loop system behaves in a satisfactory manner.

The inverse Nyquist array of the open-loop system $G(s)$ is shown in figure 9.10 for a frequency range $0 \leq \omega \leq 2$ rad/s; it is clearly seen that $\hat{G}(s)$ is diagonal domin-

Figure 9.10

it. It may be possible to improve this dominance by the choice of a constant matrix K. First choose $K = G(0)/[\det G(0)]^{\frac{1}{2}}$, that is

$$\hat{K} = \frac{1}{(0.4375)^{\frac{1}{2}}} \begin{bmatrix} 1 & 0.75 \\ 0.75 & 1 \end{bmatrix} = \begin{bmatrix} 1.51 & 1.13 \\ 1.13 & 1.51 \end{bmatrix}$$

he inverse Nyquist array for $\hat{K}\hat{G}(s)$ was plotted and it was found that, although 1e diagonal dominance at zero frequency was much improved, the dominance at gher frequencies showed marked deterioration.

Now choose
$$\hat{K} = |G(s)|_{s=j1} \left[\det |G(s)|_{s=j1}\right]^{-\frac{1}{2}}$$
that is
$$\hat{K} = \begin{bmatrix} 0.98 & 0.46 \\ 0.46 & 1.24 \end{bmatrix}$$
Therefore

$$\hat{K}\hat{G}(s) = \frac{2.29(2s+1)}{(5.73s+1)(1.48s+1)}$$

$$\begin{bmatrix} 0.98 & 0.46 \\ 0.46 & 1.24 \end{bmatrix} \begin{bmatrix} (s+1)(2s+1) & -0.75(0.5s+1)(s+1) \\ -0.75(0.5s+1)(s+1) & (0.5s+1)(2s+1) \end{bmatrix}$$

$$= \frac{2.29(2s+1)}{(5.73s+1)(1.48s+1)}$$

$$\begin{bmatrix} 0.635(s+1)(2.82s+1) & 0.275(0.5s+1)(0.673s-1) \\ 0.47(s+1)(0.986s-1) & 0.9(0.5s+1)(2.39s+1) \end{bmatrix}$$

and the inverse Nyquist array is shown plotted in figure 9.11.

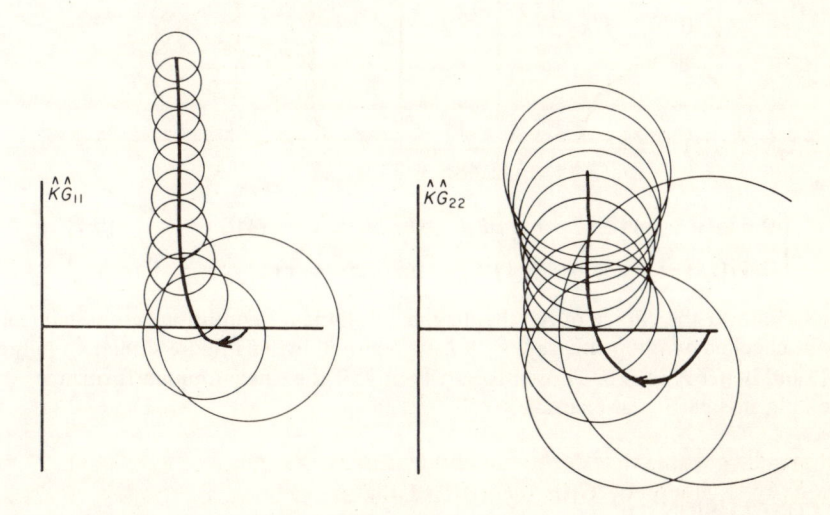

Figure 9.11

It is seen from figure 9.11 that the diagonal dominance of the system has been improved and the technique could be continued by choosing other values of $\hat{K}$ and investigating their effect on the dominance of the system.

Thus the resulting open-loop-system transfer matrix $Q(s) = G(s)K(s)$ is

$$G(s)K(s) = \begin{bmatrix} \dfrac{1}{s+1} & \dfrac{0.75}{2s+1} \\ \dfrac{0.75}{2s+1} & \dfrac{1}{0.5s+1} \end{bmatrix} \begin{bmatrix} 1.24 & -0.46 \\ -0.46 & 0.98 \end{bmatrix}$$

$$= \begin{bmatrix} \dfrac{0.89(2.39s+1)}{(s+1)(2s+1)} & -\dfrac{0.28(0.64s-1)}{(s+1)(2s+1)} \\ -\dfrac{0.47(0.96s-1)}{(2s+1)(0.5s+1)} & \dfrac{0.63(2.84s+1)}{(2s+1)(0.5s+1)} \end{bmatrix}$$

All that remains is to choose the independent single-loop controllers $k_1(s)$ and $k_2(s)$ by considering the diagonal dominant system to be two independent systems with transfer functions $Q_{11}(s)$ and $Q_{22}(s)$.

If

$$K_i(s) = \begin{bmatrix} k_1 & 0 \\ 0 & k_2 \end{bmatrix}$$

that is, a diagonal matrix of gains, then

$$\hat{K}_i = \begin{bmatrix} 1/k_1 & 0 \\ 0 & 1/k_2 \end{bmatrix}$$

and

$$\hat{K}_i \hat{K}(s) \hat{G}(s) = \dfrac{2.29(2s+1)}{(5.73s+1)(1.48s+1)}$$

$$\begin{bmatrix} 0.635(s+1)(2.82s+1)/k_1 & 0.275(0.5s+1)(0.67s-1)/k_1 \\ 0.47(s+1)(0.968s-1)/k_2 & 0.9(0.5s+1)(2.39s+1)/k_2 \end{bmatrix}$$

Thus, whatever the choice of K_i, the diagonal-dominance condition is retained (this may be checked by choosing, say, $k_1 = k_2 = 5$ and drawing a figure similar to figure 9.11) and hence K_i can be chosen to satisfy any of the single-loop performance criteria in this particular example.

9.3 CONCLUSION

The aim of this chapter has been to give an introduction to multivariable frequency-response methods. Obviously for a detailed understanding of the methods, further reading is required and a list of relevant references is given. A number of references dealing with optimal control theory is also given.

REFERENCES

MacFarlane, A. G. J., 'Return-difference and Return-ratio Matrices and their use in the Analysis and Design of Multivariable Feedback Control Systems', *Proc. I.E.E.*, 118 (1971) 2037-49.
Munro, N., 'Multivariable Systems Design Using the Inverse Nyquist Array', *Comput. Aided Des.*, 4 (1974) 222-7.
Noton, A. R. M., *Introduction to Variational Methods in Control Engineering* (Pergamon, Oxford, 1965).
Prime, H., *Modern Concepts in Control Engineering* (McGraw-Hill, New York, 1969).
Ogata, K., *Modern Control Engineering* (Prentice-Hall, Englewood Cliffs, N. J., 1970).
Rosenbrock, H. H., 'Design of Multivariable Control Systems Using the Inverse Nyquist Array', *Proc. I.E.E.*, 116 (1969) 1929-36.
——'Progress in the Design of Multivariable Control Systems', *Measurement and Control*, 4 (1971) 9-11.
——*Computer-Aided Control System Design* (Academic Press, London, 1974).

PROBLEMS

9.1 Find a matrix $K(s)$ such that $\hat{K}(s)\hat{G}(s)$ is diagonal dominant, where

$$G(s) = \begin{bmatrix} \dfrac{1}{(s+1)^2} & \dfrac{1}{(s+1)(s+2)} \\ \dfrac{1}{(s+1)(s+2)} & \dfrac{s+3}{(s+2)^2} \end{bmatrix}$$

9.2 Investigate the existence of a constant matrix K such that $\hat{K}\hat{G}(s)$ is diagonal dominant, where $G(s)$ is the transfer-function matrix of example 9.4.

9.3 Choose $K(s)$ to make $\hat{K}(s)\hat{G}(s)$ diagonal dominant, where

$$G(s) = \begin{bmatrix} \dfrac{1}{s+1} & \dfrac{1}{2s+1} \\ \dfrac{1}{2s+1} & \dfrac{1}{0.5s+1} \end{bmatrix}$$

and design single-loop controllers $k_1(s)$ and $k_2(s)$ such that the closed-loop system has a satisfactory performance.

9.4 Design a closed-loop system for the system with transfer-function matrix

$$G(s) = \begin{bmatrix} \dfrac{23.2s^2 - 354s + 13\,598}{s^3 + 53.2s^2 + 1634s + 15\,333} & \dfrac{-77.6s^2 + 817s - 17\,932}{s^3 + 125.6s^2 + 2334s + 20\,578} \\ \dfrac{0.0025s^2 + 1.18s + 33.8}{s^3 + 100.3s^2 + 2499s + 33\,573} & \dfrac{0.0055s^2 - 7.75s + 195.4}{s^3 + 67.2s^2 + 1634s + 2982} \end{bmatrix}$$

Appendix   Table of Laplace-transform pairs

$f(t)$	$f(s)$
$\delta(t)$, unit impulse at $t = 0$	1
a, a constant or step of magnitude a at $t = 0$	$\dfrac{a}{s}$
t, a ramp function	$\dfrac{1}{s^2}$
e^{-at}, an exponential function	$\dfrac{1}{s+a}$
$\sin \omega t$, a sine function	$\dfrac{\omega}{s^2 + \omega^2}$
$\cos \omega t$, a cosine function	$\dfrac{s}{s^2 + \omega^2}$
t^n	$\dfrac{n!}{s^{n+1}}$
te^{-at}	$\dfrac{1}{(s+a)^2}$
$t^n e^{-at}$	$\dfrac{n!}{(s+a)^{n+1}}$
$\dfrac{e^{-at} - e^{-bt}}{b - a}$	$\dfrac{1}{(s+a)(s+b)}$
$\sin(\omega t + \alpha)$	$\dfrac{s \sin \alpha + \omega \cos \alpha}{s^2 + \omega^2}$

$\cos(\omega t + \alpha)$	$\dfrac{s \cos \alpha - \omega \sin \alpha}{s^2 + \omega^2}$
$e^{-at} \sin \omega t$	$\dfrac{\omega}{(s + a)^2 + \omega^2}$
$e^{-at} \cos \omega t$	$\dfrac{s + a}{(s + a)^2 + \omega^2}$

Index

Absolute stability, method of Routh 75
Acceleration error constant 83
Algebraic equations, properties 32
Angle criterion 110
Asymptotic approximation 147
Auxiliary equation 25

Bandwidth 168, 172, 192
Block diagram representation 57
    algebra 60
    closed-loop control system 58
Bode diagram 129, 136, 145
    asymptotic approximation 147
    differentiating term 146
    first-order lag term 146
    first-order lead term 147
    gain margin 162
    gain term 145
    integrating term 145
    phase margin 162
    second-order term 147, 149
    use in system synthesis 192
Boiler-turbine unit 49
Break frequency 147

Characteristic equation 24
Chemical systems 16
Class of a system 81
Classical method of solution of linear differential equations 24
Classification of systems 79
    class 81
    order 81
    rank 81
Closed-loop control, definition 49

Closed-loop performance 163
    Hall chart 167
    M and N circles 165
    Nichols chart 167
    requirements 171
    transient response from frequency response 163
Closed-loop system, definition 1
    historical development 2
Complementary function 24
Computer graphics 7, 101, 106, 124, 207, 211
Conditional stability 158
Conformal mapping 155, 163
Conservation of mass, energy and momentum 16
Control system synthesis *see* Synthesis of control systems
Controller gain 73
Conversational program 127
Convolution integral 55
Corner frequency 147
Cut-off frequency *see* Bandwidth

Damped frequency 45, 140
Damping factor 41, 89, 140, 171
Derivative action 93
Diagonal dominance 216
    achievement of 218
Differential equations, second-order 41
Disturbances 60, 95
    steady-state error 86

Electrical systems 12
Encirclement theorem 155

Index

Feedforward control 205
First-order lag 87
Forward transference 81
Frequency, break 147
   corner 147
   cut-off 168, 172, 192
   damped 45, 140
   gain crossover 192
   natural 44, 91, 140
   of oscillation 91, 140
   resonant 140, 171
   undamped natural 41, 89, 140
Frequency domain 129
Frequency of oscillation 91, 140
Frequency response 129
   derivation 130
   first-order system 132
   graphical representation 136
   interpretation in s-plane 134
   second-order system 132
   system synthesis 192

Gain crossover frequency 192
Gain margin 161
   Bode diagram 161
   Nichols diagram 162
   Nyquist diagram 161
Governor system 98
Graphics terminal see Computer graphics

Hall chart 167

Impulse function 56
Integral action 95
Interaction 211
   diagonal dominance 216
   measure of 218
Inverse-Nyquist-array technique 211, 215, 216
   diagonal dominance 216
     achievement of 218
   stability theorems 216
Inverse Nyquist diagram 172
   stability criterion 172
   system synthesis 203
Inverted pendulum

Kirchhoff's current and voltage laws 13

Lag–lead network compensation 183
Lag-network compensation 180, 187

Laplace transformation, definition 27
   differentiation 28
   exponential function 31
   exponentially decaying sine function 31
   impulse function 30
   integration 29
   inverse 27
   linearity 28
   ramp function 31
   sine function 31
   solution of linear differential equations 41
   s-plane translation 30
   step function 31
   superposition 28
   table 226
   time translation 29
Lead-network compensation 177, 187
Linear differential equations, solution, classical method 24
   Laplace transform method 27, 39
Linearisation 18

M and N circles 165
   Hall chart 167
   inverse Nyquist diagram 173
   Nichols chart 169
   system synthesis 198
Magnitude criterion 110
Mason's rule 66
Mathematical model 5, 39, 53
Mathematical modelling 4, 58
Maximum amplitude 140, 171
Mechanical systems 5
Multivariable systems 210, 212, 213

Natural frequency 44, 91, 140
Newton's second law of motion 6
Nichols chart 169, 200
Nichols diagram 129, 136, 153
   gain margin 162
   phase margin 162
   second-order lag 153
Non-minimum phase systems 144
Nyquist diagram 129, 136, 137
   differentiating term 138
   first-order term 139
   gain margin 161
   gain term 138
   integrating term 138
   phase margin 161

Index

Nyquist diagram (*cont.*)
   second-order term 139
   stability 154
   system synthesis 198
Nyquist's stability criterion 129, 157, 158
   inverse Nyquist diagram 172

Open-loop control 49
Open-loop system 1
Optimal control theory 211
Order of a system 81

Parallel compensation 184
Partial fraction expansion 35
   equating coefficients 37
   residues 36, 38
Particular integral 24
Percentage overshoot 45, 91, 171
Performance criteria 73
   bandwidth 168
   second-order system 90
      natural frequency 91
      percentage overshoot 91
      predominant time constant 91
      rise time 91
      settling time 91
   stability 74
Performance requirements, bandwidth 172
   damping factor 171
   maximum amplitude 171
   percentage overshoot 171
   resonant frequency 171
Phase margin 161
   Bode diagram 162
   Nichols diagram 162
   Nyquist diagram 161
Polar form 107
Polar plot 129, 136, 137
Poles 55, 106
Pole–zero configuration 106
Polynomials, roots of 32
   even order 34
   odd order 33
Position error constant 82
Predominant time constant 91
Pure time delay, transfer function 54

Rank of a system 81
Regulator 2, 52, 60
Relative stability 161

Resonant frequency 140, 171
Rise time 91
Root-locus method 105
   angle criterion 110
   definition 109
   determination of transient response 120
   magnitude criterion 110
   rules for drawing 111
   system synthesis 186
Roots of polynomials 32
   even order 34
   odd order 33
Routh's method 75
   all-zero row 78
   zero in first column 77
Rules for drawing root-locus diagram 111

Second-order differential equations 41
Second-order systems 89
Sensitivity of control systems 96, 172
Series compensation 177
   lag–lead network 183
   lag network 180
   lead network 177
   variable gain 177
Servomechanisms 2, 52
   historical development 2
Settling time 91
Signal flow graphs 64
   Mason's rule 64
Single-input single-output (SISO) systems 210
Stability, absolute 75
   conditional 158
   examples of 144, 153
   heuristic approach 134
   Nyquist diagram 154
   pole–zero configuration 106
   roots 27, 39
   transfer function 55
Steady-state analysis 79
Steady-state error, acceleration error constant 83
   disturbances 86
   position error constant 82
   table of 84
   velocity error constant 83
Steady-state gain 55
Synthesis of control systems 175
   Bode diagram 192

Synthesis of control systems (*cont.*)
    feedforward control 205
    frequency response method 192
    inverse Nyquist diagram 203
    M and N circles 198
    Nichols chart 200
    Nyquist diagram 198
    parallel compensation 184
    root-locus method 186
    series compensation 177

Table of Laplace transforms 226
Tachometer 184, 204
Time constant 8, 25, 87
    predominant 91
Time domain 73
Trailer suspension system model 9
Transfer function 53
    definition 54
    pure time delay 54
    stability 55
    steady-state gain 55
Transfer function matrix 212
    controller 216, 218
    inverse 212

Transient behaviour of control systems 87
    addition of velocity feedback 87
    derivative action 93
    effect of disturbances, integral action 95
    first-order systems 87
    second-order systems 89
Transient response 87
    determination from frequency response 163
    determination from root-locus diagram 120

Undamped natural frequency 41, 89, 140

Variable gain compensation 177
Velocity error constant 83
Velocity feedback 92, 184, 204

Weighting function 56

Zeros 55, 106